P9-DHQ-879

HISTORY, FOLKLORE, ANCIENT RECIPES

ARNEO NIZZOLI

THE SQUASH

CELEBRATED IN RECIPES FROM ANCIENT TIMES TO THE PRESENT
INTRODUCTION BY ALBERTO CAPATTI
ACCOMPANYING WINES SELECTED BY GIUSEPPE VACCARINI

KÖNEMANN

All rights reserved. Rights of translation, electronic recording, reproduction and adaptation, in whole or in part, by whatsoever means (including microfilm and photocopies) are reserved in all countries and may not be granted to third parties without written permission from the publisher.

COPYRIGHT © 1996 BIBLIOTHECA CULINARIA S.R.L., LODI

GRAPHIC AND EDITORIAL COORDINATION: MARIO CUCCI
COMPILATION AND TEXTS: DANIELA GARAVINI
GRAPHICS AND DESIGN: THE C' SERVIZI EDITORIALI SAS, MILAN
PHOTOGRAPHS: NICOLETTA INNOCENTI
OTHER ILLUSTRATIONS: "ACHILLE BERTARELLI" PRINT COLLECTION, NATURAL HISTORY LIBRARY, MILAN

COPYRIGHT © 1998 FOR THE ENGLISH-LANGUAGE EDITION
KÖNEMANN VERLAGSGESELLSCHAFT MBH
BONNER STR. 126, D-50968 KÖLN

ENGLISH TRANSLATION: PAMELA MARWOOD/ROS SCHWARTZ TRANSLATIONS, LONDON
EDITING/TYPESETTING: GRAPEVINE PUBLISHING SERVICES, LONDON
PRODUCTION MANAGER: DETLEV SCHAPER
ASSISTANT: NICOLA LEURS
PRINTING AND BINDING: KOSSUTH PRINTING HOUSE CO.

PRINTED IN HUNGARY

ISBN 3-8290-1462-7

10 9 8 7 6 5 4 3 2 1

Contents

A quirk of nature

by Alberto Capatti

The first man to eat squash was Marullo Egizio.
ORTENSIO LANDO, COMMENTARY

The squash, with its 90 genera and 700 species and in its great variety of shapes and sizes, weights and colors, epitomizes the excesses and quirks of Nature. Its origins as a food reach back to the dawn of history: in Europe starting with the gourds and other cucurbits of the ancient Mediterranean region and in the Americas with the squashes cultivated by the forbears of the Inca and Aztec. It is now found all over the world, from the United States to Africa and Asia. Its importance, far from being limited to its nutritional value, extends into the spheres of domestic ornament, art, language and celebrations. Squashes can be white or yellow, soft or dried; they can be hollowed out to make floats for swimmers or fishnets, flasks to hold wine, or containers in which peasants keep salt or that fishermen use for their catch; with a few seeds left inside, the hollow squash becomes a rattle. In Les Halles, the meat and vegetable market in Paris, a feast known as *la fête du potiron* was held during the month of September in honor of King Pumpkin; paintings record its ornamental, animistic and erotic values (there is, for example, a still life by Juan

C. Crivelli, *Virgin and Child*
Ancona, Pinacoteca Comunale Podesti

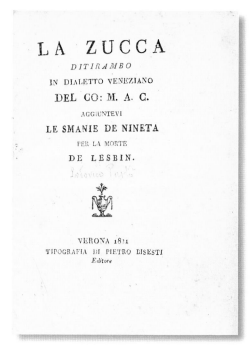

*La zucca: Dithyramb in Venetian dialect,
Verona 1824*

Sanchez Cotán, a Virgin and Child by Crivelli, and scenes of the Temptation of Adam by Ferdinando Maria Campani)[1]. As the largest of all fruits, the squash has never ceased to arouse wonder: the *Guinness Book of Records* for 1996 cites the squash from Ashton in Canada, which weighed 988lbs/449kg.

Although it is large and bulbous and can take an almost infinite variety of shapes, the squash is a vegetable that has its own unique characteristics and straightforward uses. In Rome, from the 1st century BC to the 3rd AD, it figured in simple meals, and was valued in the same ways as the water melon; it was considered healthy and was prepared in many different ways. Apicius gives nine recipes: the squash is first boiled, then flavored with spices and condiments, or accompanied by a sauce; or it may be fried, cut up or mashed, or perhaps mixed with chicken. Its appearance in mean or poor meals, as recorded in the writing of Martial and Pliny, heralds its fortune in the modern age. It is also universally used as a container[2].

The eccentric, oblong *zucca* figures in horticultural manuals and all Italian recipe books from the anonymous Tuscan author of the late 14th century right up to Pellegrino Artusi in the late 19th, while *Cucurbita maxima*, the round, flattened, heavy squash, became known in Europe through the exploits of the earliest explorers of the New World; it appeared in their botanic atlases, arousing fresh interest in all indigenous species of cucurbits. The pumpkin has figured in fables since the early 16th century, even appearing in humorous texts.

For a vegetable to feature so prominently in poetry and stories, it must have aroused universal interest, attracting attention and commanding a certain respect. There are few fruits that writers succeed in cultivating in the field of literature. The cucurbit (be it the gourd, squash, melon or pumpkin) enjoys this privilege. It seems out of proportion, and so features in humorous or satirical texts; and it is humble, and so fits comfortably with vernacular dialogue. Because it can grow to full size in six to eight days, it embodies speed but also heaviness. Because it has been cultivated for millennia, it is part of human history.

---

1 For an exhaustive, multidisciplinary treatment of the squash, see:
  Ralf Norman and Jon Haarberg, *Nature and language. A semiotic study of cocurbits in literature*, London, Routledge & Kegan Paul, 1980

2 J. André, *L'alimentation et la cuisine à Rome*, Paris, Les belles lettres 1981, pp.41–42

As an epithet for the head of the house, the pumpkin has been transmitted from Greek and Latin aphorisms to many modern languages. It is almost as if the pumpkin were destined to grow not only in volume but also in fame. Its fulsome presence can be observed in the Italian Renaissance, reflecting the cultural florescence of this splendid period. Its various shapes may suggest the swollen belly of a pregnant woman or an obscene penis; its use as a rattle or a bottle conjures up associations with music and drinking. Teofilo Folengo, the 16th-century Italian satirical writer, fancifully describing poets, singers, astrologers and other characters enclosed within an enormous empty pumpkin, concludes his poem with this bantering statement: *"Zucca mihi patria est"*[3] ("The pumpkin is my motherland"). The huge head of the vegetable family becomes a ridiculous analogy for the intellect.

This Renaissance jester was not alone in fêting the pumpkin. In 1543, in Piacenza, the academy of the Ortolani (market gardeners) was formed, its members being scholars who took names such as Cetriolo (cucumber), Popone (melon) and Cocomero (water melon), while Francesco Doni, the founder, chose to be known as Semenza (seed). This individual was a roaming Florentine, the flamboyant author of *La zucca*, a satirical work stuffed with chit chat, anecdotes and information, an ideal source for dinner table talk. In the greeting to his readers Doni actually began with a message to the kitchen: *"I want to leave the prepared dish of pumpkin, the boiled squash with egg, stewed in unripe grape juice, in a dip with mixed spices, fried with the sauce, that is, roast …"*[4].

The principal merit of this eccentric work lies in its listing of recipes with one common ingredient, the squash (a food available in quantity, fresh or dried, easy to cook and easy to flavor). Costanzo Felici, a natural historian from Rimini and a contemporary of Doni, in a letter "Of salads and plants that in whatsoever way come to feed mankind", writes in the same vein: *"The squash … is habitually oft in men's food, cooked in soup, in cakes, in fritters, with meat, with oil, with cheese, with egg and in many ways, as best can be devised by good cooks"*[5].

These observers inscribe squash in the domestic calendar: "As soon as the squash begins to run its arms across the ground, it is ready for the kitchen"[6]. By virtue of its abundance and variety and its vigorous growth, it was food right on the doorstep, for everyone, with whatever condiment their fortune allowed, from salt to sugar, and however it was cooked: lightly boiled for use in salads, and as long as possible for soups and purées.

Quite different, however, was its role in courtly circles in the Lombardy Plain, particularly at the court of Ferrara; writing in the 16th century, Cristoforo di Messisburgo and Rossetti describe how squashes were filled with pheasant and chicken, or whole pigeons boned and stuffed; alternatively, seasoned with cheese

3 Teofilo Folengo, *Baldus,* edited by Emilio Faccioli, Turin, Einaudi 1989, p.878

4 Anton Francesco Doni, *La Zucca*, in: Works by Pietro Aretino and Anton Francesco Doni, Naples, Ricciardi 1976, p.604

5 Costanzo Felici, *Dell'insalata e piante che in qualunque modo vengono per cibo dell'homo*, Urbino, Quattroventi, p.88

6 Vincenzo Tanara, *L'economia del cittadino in villa*, Venice 1680, p.258

it made filling for *tortelli*, or was used to cover capon meat. No record exists of squash-based recipes for royal tables because it was served as a side-dish or accompaniment to richer or more highly prized foods. It is likewise omnipresent with luscious meats and appetizing flavorings such as sugar or saffron. The verdict of Io Stefani, a gifted cook from Bologna, was undoubtedly instrumental in its rise from the status of modest vegetable to great culinary heights. In 1642 he stated: "*It can be made into such a variety of dishes that it forms an entire meal*"[7].

Its role as a serving vessel at important dinners deserves special mention. Bartolomeo Scappi[8] gives us to understand that, for a cook, stuffing a squash is no problem: its capacity, in fact, allows it to be filled with pounded pork or veal with eggs and ham, or layers of yellow brains and filleted chickens or little birds. The difficulties arise in the cooking, during which its shape and texture must be preserved. Two methods are described, for

Handwritten letter from a certain Sella, unidentified, addressed "To the most highly esteemed Signor Francesco, cook of the Galliani household" with a description of a recipe for fried squash.

cooking the whole squash either in a saucepan or in the oven. The squash, hollowed out and stuffed, is boiled in fatty broth with pepper, cinnamon and saffron "*until it acquires flavor and is not insipid*". Transferring it from the cauldron to a large dish "*in such a way as to serve it hot surrounded by the ham or the pork belly*" must have been a feat of dexterity. The alternative, in the oven, demands no lesser ability. The squash is lined with slices of ham and packed with all sorts of good things; the hole is then stopped with squash flesh and the whole secured with

string. It is then wrapped in sheets of paper and placed "*on a copper or clay base*", in an oven "*less hot than if you were going to cook bread*" for two hours. Taking it out without breaking it, removing the paper and string, and decorating it with leaves was quite an art.

In these recipes the squash regains its magical powers: it is a stomach bursting with savory food, hiding a surprise within its colorful rind and transforming the table into a divine vegetable garden. This sort of culinary prowess did not disappear at the end of the 16th century and was indeed not confined to Europe; this is not surprising given the great adaptability of the squash in a legion of different recipes regardless of historical period and geographical location. In The *Life and Passion of a Chinese Gastronome* by Lu Wenfu[9], one of the most delicious novels to come out of contemporary China, the "squash surprise" plays a providential role during the famines of the years 1959–61. The gastronome Zhu Ziye considers his meager resources, hunts

7 Bartolomeo Stefani, *L'arte di ben cucinare*, Mantova 1662, p.84
8 Bartolomeo Scappi, *Opera*, Venice 1570, p.82
9 Lu Wenfu, *Vita e passione di un gastronomo cinese*, Parma, Guanda 1991, p.82

through his archives and devises the following recipe, which he describes thus to his accomplice and rival Gao:
"*We could invent the squash surprise, fill it with high-quality rice, add eight 'treasures' and steam it; by uniting the freshness of the squash, the sweetness of the gluten and the eight 'treasures' we would achieve a harmonious combination*"[9].

The fact that the squash is an inexpensive food is doubtless at the root of these inventions, either because it can be stuffed with more costly choice morsels or because, by its volume, it produces the illusion of abundance. The luxuriant growth of the squash in any sort of soil and in hundreds of shapes is the cause of fascination and has induced minutely detailed observation of its metamorphosis. The squash that appear in Giuseppe Arcimboldo's portrait of *Rudolph II as Autumn* conjure up thoughts of all manner of dishes: the tops and the tender young zucchini for salad, the rinds in

compote, the flowers stuffed, floured, coated in batter and fried, and the seeds as well, all play a part in the gastronomic spectacle. Squash also feature in rustic pharmacopoeia; the seeds themselves purify the kidneys, while the flowers soaked in oil "*for a summer in the sun*" serve to refresh them[10].

After its heyday during the Renaissance and Baroque periods, the squash

Arcimboldo, Portrait of Rudolph II as Autumn, Vienna c.1590

continued to be seen on courtly tables, albeit with some modification of tone and a lesser interest in its dramatic possibilities. Corrado saw fit to use it in Naples in 1773 and Vialardi, in the service of Carlo Alberto and Victor Emmanuel II, suggests it in a treatise of 1853. What cruel fate swept away the squash in more recent times? We seek it in vain in Auguste Escoffier's *Guide Culinaire* (1903), bible of professional cooks the world over and the model for many subsequent recipe books. Mario Borrini[11] devotes to the squash four recipes and a homemade soup, Il Carnacina. On the domestic front its retreat into obscurity was signaled by Artusi, who in 1891 mentions it but twice, in a soup and in a cake[12]. This is symptomatic of a rift between the high table, the urban kitchen and the rural board; between modern culinary practice in the big hotels and time-honored, deeply rooted tradition and popular custom that springs from the vegetable garden and is inspired by it.

10 Vincenzo Tanara, op. cit., p.259

11 Mario Borrini, *La cucina pratica professionale*, Novara, Borrini, 1960, p 484

12 Pellegrino Artusi, *La scienza in cucina*, Turin, Einaudi, 1970, p.563

Were it not for its astonishingly rapid growth the squash would not appear particularly interesting. It is certainly not ideal for soufflés and gratins, puddings and cream puffs. It has been supplanted for flavor by the zucchino, and for texture by the potato. At most a small slice of squash might be added to soup.

However, the squash is returning to favor and prominence. Whereas international French cuisine banished the squash, the revival of old recipes has reinstated it. While the food industry has devoted its energies to carrying out research on the potato, developing instant mashed potato and frozen chips, the squash (the oldest inhabitant of the vegetable garden) has escaped the ravages of the icebox and the can. It is thanks to our broadening horizons, brought about by the import and export of foreign and homegrown fruit and vegetables, and the quest for color and flavor, that the squash and its relatives have been rehabilitated, reappearing as novel vegetables despite their ancient origins. Their new-found novelty enhances their nobility, which is in turn taken to new heights by the recipes that follow in this book. Vegetables so closely form part of human history that concepts of what is ancient, old, or modern are purely relative: in America, no Thanksgiving celebration would be complete without the relatively recent festive symbol of pumpkin pie. It is not inconceivable that the squash will be reborn, aristocratic and astonishing, the symbol of inexhaustible mastery and a central part of folk culture. If its progress on the table is as fast as in the fields, we shall eat nothing else.

Pietro Andrea Mattioli.
Commentarii in sex libros
Redacii Dioscoridis Anazarbeis de Medica Materia

About the squash

DANIELA GARAVINI

Names and origins

Botanical names: *Cucurbita maxima, Cucurbita moschata, Cucurbita pepo, Lagenaria vulgaris*

American Indian:	*askutasquash*
English:	*pumpkin, marrow, courgette*
French:	*courge, potiron*
German:	*Kürbis*
Italian:	*zucca*
Portuguese:	*aboboreira, cabaça*
Russian:	*tyjva*
Spanish:	*calabaza*
US English:	*squash, pumpkin, zucchini*

What is the origin of the name *cucurbita*? It may derive from the late Latin *cocuti*, meaning "head". To this day all over southern Italy the squash is known as *cucuzza* or *cocuzza*; the name *"suca"*, used further north, may be a corruption of this. The word *zucchino* is a diminutive of the latter, while *courgette* is a diminutive of the French word *courge*. Squash is thought to derive from the American Indian *askutasquash*. And where does the vegetable itself come from? It is present all over the world in great variety, and was known and used in the kitchens of ancient Rome. The yellow winter squash that we appreciate more today originated in tropical America and arrived in Europe in the 16th century. The first to become widespread in Europe was *Cucurbita pepo* (before 1542). This was followed by *Cucurbita maxima* (by 1544) and finally *Cucurbita moschata* (by 1591). The first writers to describe them were

The instruction in this country proverb is still entirely valid, at least in central and northern Italy. Squashes do in fact suffer from frost if they are sown too early (they are annual plants and so must be sown afresh every year). This does not apply in the south, where the climate is milder.

About a month after the seeds are sown , the plant has grown sufficiently to produce flowers. The first to bloom are the male flowers. Of the female ones, only a few will be fertilized and develop into fruit.

The yellow squash is gathered when fully mature; this occurs in summer or early autumn, according to variety. In Italy it is cultivated mainly in Lombardy, Emilia Romagna, and the Veneto, on the plains stretching out on either side of the Po, in Campagnia and in Puglia. Over the last 20 years, however, production has fallen considerably. Compared with the 9055 tons/9,200,000 kg produced in 1994, 25,984 tons/26,400,000 kg were harvested in 1987, and 57,185 tons/58,100,000 kg in 1976. In fact production and consumption today is rather less than a sixth of what it was about 20 years ago. The squash, like leguminous vegetables (production and consumption of which is also falling noticeably), would actually seem to be another victim of our changing lifestyle.

Pierrandrea Mattioli in 1544 and Gherardo Cobo around 1550. A witness to their cultivation in Italy was the agronomist Agostino Gallo, from Brescia, who mentions three varieties of *zucca*: "white", "marine" and "Turkish". Two others, also writing in the 16th century, speak generically of the *zucca*; one is the cook Cristoforo di Messisbugo, of Ferrara, who provides a description of the vegetable (and a recipe can be found on page 27), and the other the doctor and naturalist Costanzo Felici, of the Marches.

Cultivation

"Poni la zucca in aprile, ti verra grossa come un barile."
"Plant the squash in April and it will grow fat as a barrel."

A generous vegetable

The popularity of the squash in rural Italy up to a few decades ago was due to a combination of factors. First and foremost is its generous size and nutritional value as a foodstuff. As will be seen from the recipes that follow, all parts of the vegetable can be eaten, and those varieties that are unsuitable for the table can be fed to animals. Secondly, it is very easy to cultivate, even in poor soil, marginal ground, or areas that cannot be irrigated; indeed excessive fertilizer or watering can actually harm it. The plant is not particularly susceptible to insect infestation or diseases. In fact, if it is just sown at the right time and protected from frost, it will thrive without attention.

Thirdly, a factor not to be undervalued in flat areas under intense cultivation, is its ability to provide shade in the hottest months of the year. All that is necessary is to train the squash up specially erected poles and the plant will rapidly grow leaves that create a pergola or arbor that provides shade from the midday sun or a shelter to keep tools or animals. When the vegetables begin to form they can be supported on a sort of shelf balanced on a stake inserted in the ground.

Finally, squash keeps extremely well. Picked in late summer or early autumn, it remains perfectly edible and sound until the following spring. It does, however, need to be protected from the cold. During the winter, peasants living on the Lombardy Plain used to keep their squash in the bedroom, either under the bed or on the bedside table.

The squash plant

The squash plant has a ridged stem usually covered with rough skin. The leaves are large, palmate and lobed, with irregularly denticulate edges, and grow from a long stalk also covered with rough skin. Near the base of the leaves are the tendrils that enable the plant to climb when it finds suitable support; otherwise its habit is trailing. The flowers have a well developed corolla, which is semitubular and orange. The fruit is a *peponio*, or berry; when it is ripe, the flesh can vary in color from light shades of yellow to a rich orange. A cavity in the center of the fruit contains the seeds, which are flat, oval, and white. This is a species with a tremendously long agricultural history and a huge number of varieties that can be distinguished by shape, color, and the appearance of the rind and the flesh. Most importantly, it is very easily hybridized, thus producing an infinite number of variants. It is therefore not surprising that the nomenclature of the squash family is often contradictory, with regard both to botanical names and common ones.

In general the name *Cucurbita maxima* refers to the yellow squash, tending to be roundish in shape. *Cucurbita pepo* designates the zucchino, which has a green rind and white

flesh, and which is gathered and eaten while young and tender. *Cucurbita moschata*, the yellow musk melon, tends to have an elongated shape. The latter species, or a very similar form, is considered by scholars of paleobotany to be the primitive type of squash from which all the others have evolved. *Lagenaria siceraria* is the bottle-shaped gourd; generally it has always been hollowed out and dried for use as a container, even for liquids, although some varieties are edible provided they are picked before they are completely mature; in fact this was the *zucca* that was eaten in Europe before the arrival of the American squash.

Varieties of squash

The varieties of the species *Cucurbita maxima* give very large fruit. They are globular and may be elongated at either end. The rind may be smooth, ridged or warty and may range from dark green, yellow, or orange to grayish, or variegated.

The flesh is firm and dense, though in general not fibrous, and varies in color from yellow to orange.

The varieties of the species *Cucurbita moschata* are generally round or elongated, the rind is smooth, ranging in color from yellow to sea green, the flesh yellow to rich orange and of a dense consistency.

The species *Cucurbita pepo* includes mainly the numerous varieties of zucchini, which will not be described in this book. There are, however, some winter squashes belonging to the same species.

The varieties of the species *Lagenaria*, which are Asian and African in origin, have a mainly ornamental value, but in the past they had a practical value too, being dried and hollowed out and used as containers, especially for liquids. The shape of our bottles and flasks is undoubtedly derived from them.

The most common varieties

Marina di Chioggia (*Cucurbita maxima*): rounded shape, flattened at either end; the rind generally green, ridged and warty; the flesh orange. Much prized for its culinary value.

Piacentina (*Cucurbita maxima*): rounded shape, flattened at either end; the rind may be green or gray, generally ridged and warty; the flesh is orange, firm and floury, and slightly fibrous. Much prized for its culinary value.

Mantovana (*Cucurbita maxima*): similar in appearance to the Piacentina; flattened at either end; the rind wrinkled and ridged, and varying from green to gray; the flesh orange, doughy and sweet. Highly prized.

Mammouth (*Cucurbita maxima*): rounder than the Piacentina or Mantovana; the rind orange; the flesh yellow. Its peculiarity, as the name implies, is that it can reach gigantic proportions and weigh over 100 lb/45 kg. It has no great culinary value.

Turbante (*Cucurbita maxima*): also known as "*cappello del prete*" (priest's hat) or "*turca*" (Turkish). It has a very unusual shape, lower extremity smaller than that nearest the stalk and often a different color. Its value is mainly ornamental, but it is also good to eat.

Ungherese (*Cucurbita moschata*): elongated shape; the rind orange or beige tending towards violet; the flesh yellow. Much prized in cooking.

Violina (*Cucurbita moschata*): shaped like a violin; the rind smooth and beige; the flesh fiber-free and well flavored. Bears similarities to the cultivar Butternut Squash.

Chioggia (Cucurbita moschata): in the Veneto region it is known as *suca baruca*, "*suca*" being the local pronunciation of "*zucca*", and "*baruca*" meaning "verruca" or "wart", indicating the warty appearance of the rind. An alternative etymology for the word "*baruca*" may be a relation to the Hebrew "*baruch*", meaning "saint" or "blessed". This recalls the other name by which this zucca is known: "*zucca* *santa"* (holy zucca), which may be a vestige of Jewish influence on both the language and the food of this region, resulting from the Jewish area that once existed in Venice.

Piena di Napoli (*Cucurbita moschata*): cylindrically shaped; the rind smooth, and dark green or orange; the flesh orange with longitudinal fibers. It can grow to a length of over three feet. Much prized for its culinary value.

A tromboncino or **d'Albenga** (*Cucurbita moschata*): unlike the others this one is gathered in summer before it is ripe, when its fruits are similar to zucchini, though larger and with a thicker rind. The flesh is whitish, tender and tasty.

Lagenaria (*Lagenaria leucanthera*): can be eaten if gathered early when the fruit is no more than 10 in (25 cm) long.

Bianca or **Benincasa** (*Cucurbita pepo*): of Asian origin; cylindral or round; the fruit bristly when young, smooth and waxy when mature; the flesh floury and white. It should be picked young.

The squash as part of a healthy diet

The squash is a particularly healthy food because, as we have seen, it does not thrive on over-fertile soil and so does not run the risk of accumulating nitrates, as many other vegetables do, and because it is naturally resistant to parasitic infestation it is almost never treated with phytochemicals. Also it does not require special conservation treatment, but will keep well in anywhere where the temperature does not fall below about 45°F/8°C.

It is one of the few vegetables that has a well defined season and is not grown in greenhouses. At the most, precocious varieties are selected, which are sown first and so arrive first on the market, or traditional varieties are subjected to forcing, which means sowing them a month early, under plastic to protect them from frost, so that they provide early fruit.

Vegetable spaghetti

The squash known as "vegetable spaghetti" is in great demand in the United States because of its lightness. The flesh looks like a mass of spaghetti and it can be eaten stewed (the ripe fruit must boil for 40 minutes) either as a vegetable or as a kind of pasta, to be dressed to taste. While the seeds are widely available, the fruit can only rarely be found, and then as a curiosity.

The squash is both low in calories (averaging about 17 calories per 4 oz/100 g) and very filling and so is ideal as part of a slimming diet. It has a high fiber content and is rich in vitamin A, potassium and other substances with antioxidant (and therefore anti-aging) properties. The yellow squash is one of the richest natural sources of vitamin A: 7 oz/200 g of yellow squash provide the average daily requirement. Vitamin A plays a multiple role in the human organism: it contributes to the formation of all the epithelial tissues (skin and mucous tissue) and plays an important role in cell renewal. As it is a powerful anti-oxidant it inhibits the formation of free radicals, thus combating cell pollution, premature aging and tissue degeneration, and also provides some protection against cancer.

In the fall and during winter the squash can amply replace vegetables such as carrots, broccoli or spinach as a source of vitamin A. The yellow squash is second only to the potato among the foods most rich in potassium. The seeds of the squash are also highly nutritional; they comprise 18.7% protein, 50.5% fat, 24% carbohydrate and 5.6% fiber, and also contain traces of iron, zinc, phosphorus and vitamins.

Medicinal uses

The squash has long been known for its refreshing and emollient properties, and for its use as a laxative, diuretic and sedative, as a tonic and to combat infections of the digestive tract and remedy disorders of the heart and kidneys.

The best-documented medicinal use of the squash is as a cure for such acute infections of the digestive system as enteritis, dysentery and typhoid fever. It is said that, during the cholera epidemic in Mantua in the early 20th century, many people put their faith in the popular tradition that prescribed diets and decoctions of squash flesh. The active elements in this case were very probably the squash's high concentration of potassium and magnesium, which rapidly helps to replenish with minerals a debilitated and dehydrated organism, and the emollient effect of squash on the digestive tract.

Traditional Chinese medicine also recognized the diuretic, cleansing and anti-inflammatory properties of squash (particularly *Cucurbita moschata*) and also its soothing effect on those suffering from coughs and bronchial asthma.

The main therapeutic use of squash seeds is as a vermifuge, mainly to flush out tapeworm. The seeds' action depends on an amino acid, cucurbitina, which paralyses the worms and dislodges them from the walls of the intestine. The advantage of squash seeds over other more powerful

Average chemical composition of the yellow squash
(percentage of edible part)

WATER	PROTEIN	LIPIDS	GLYCIDS	FIBER
94.6%	1.1%	0.1%	2.7%	0.8%

CALCIUM	PHOSPHORUS	IRON	SODIUM	POTASSIUM
0.03%	0.04%	0.0007%	0.007%	0.35%

VIT. B1	VIT. B2	VIT. P	VIT. A	VIT. C
0.00003%	0.0006%	0.00182%	0.28%	0.013%

17 calories per 100 g

vermifuges is their complete harmlessness, so that they can be confidently given to children until the desired effect has been achieved.

Popular tradition attributes a further virtue to squash seeds: they are said to be a potent aphrodisiac.

Every part is edible

There is one characteristic of the squash that it shares with few other vegetables. Every part of the plant can be eaten, including the leaves and tender shoots, which can be cooked in omelets or made into soup.

The flowers are considerably more appetizing. Both the male flowers and the female, which are larger, can be eaten. The flower of the squash is distinct from that of the zucchino in that it grows on a stalk. They are excellent when dipped in batter and fried in very hot oil, particularly when filled with a piece of desalinated anchovy, or mozzarella cheese, or a heavier mixture. They can also be eaten in salad or added to a risotto just before the rice is cooked, or to a vegetable soup. Whatever cooking method is used, it is important to make sure that they are absolutely fresh, with no sign of withering.

As to the flesh, this can obviously be prepared in a great variety of ways, as the recipes in this book demonstrate. The seeds, toasted and lightly salted, were once a widely eaten snack. Known mainly as *bruscolini*, they were a typical snack for the cinema or circus. Today the habit has lost out to other, richer, products. But wouldn't you sometimes like to eat a few *bruscolini* instead of popcorn, potato chips or other savory snacks?

Even the rind of the squash can be eaten. It forms the main component of the recipe described by Cristoforo di Messisbugo on page 27 and that by Arneo Nizzoli given on page 37. In southern Italy, particularly in Calabria, the rind of the squash, or rather the white part of it, like that of melons and water melons, would be cut into strips and hung by a cord in the window to dry in the summer sun.

In days when food was scarce the strips of rind would be soaked in water, then boiled in salted water, sprinkled with flour and fried. As a means of laying in food to fall back upon in lean times, the habit of drying squash and melon rinds has been lost, and even eating them marinated is now merely a gastronomic curiosity. This gives an idea, however, as to how well the squash lent itself to the art of making the maximum use of all resources, which has always

characterized rural civilizations. Whatever little remains of the squash burns well in the fireplace!

Buying squash

The first squashes arrive on the Italian market in June. They are early varieties, which produce medium-sized, roundish fruit with sweet, yellow, fragrant flesh.

The summer varieties grown in southern Italy, particularly in Campania and Puglia, are ready in July. The most famous variety is known as *zucca di Napoli* (Neapolitan zucca). Because of its enormous dimensions (it can measure over 3 ft/1 m), it is inadvisable to buy a whole one. It can be bought in slices, best when cut on the spot. At the supermarket one can check the date it was prepared. When buying precut slices, check that the rind is whole and sound, without spots of rot, and that the seeds have not turned black. If they have, select another.

Toward the end of summer the squash typical of northern Italy – the Marina di Chioggia and the Piacentina – arrive on the market. They can be bought whole (their average weight is 3–11 lbs/3–5 kilos), or else cut. If you are buying them whole, check that the rind is unblemished, sound, and without soft or rotten spots. If you buy them cut, make sure that the flesh is firm and floury. The thickness of the flesh is also an indication of the quality of the squash.

The color of the flesh should be between a rich yellow and orange. Do not buy a squash if the flesh is slightly green. If you are on good terms with the shopkeeper ask him to let you taste a slice: good squash taste good raw!

Basic procedures

From the gastronomic point of view, there are two major types of squash: one with firm, compact flesh, with a decidedly sugary flavor; the other with more fibrous flesh and a more watery flavor. The first, the best known of which is the *zucca marina*, just like the *zucca piacentina*, lends itself well to the preparation of purées or stuffings, and to being fried in exactly the same way as potatoes. The second type, of which the best example is the *zucca piena* from Naples, is better used in soups, mixed sauces or marinades. Very many squash recipes involve the preliminary cooking of the squash, and in general one of the seven following methods is used. Cooked then whisked or sieved, the flesh provides the base for a simple purée or filling for *tortelli*, savory pies, gnocchi, soufflés, sweet dessert molds and so on.

Boiling

Cut the squash into evenly sized pieces and add to boiling salted water, checking during cooking that they do not disintegrate. The cooking water will be rich in mineral salts and can be used for soups.

USEFUL TIPS

⩚ Cutting a whole squash with a knife is a rather arduous task. The best way to open it is to drop it on the ground (in the yard, though, not on the kitchen floor!) and it will split in half.

⩚ If the flesh is to be mashed, remember that it is much quicker to remove the rind once the squash is cooked.

⩚ If the flesh is to be diced, the rind must be removed from the squash while raw. It is easiest to do this by cutting the squash into slices, not too thick, laying them flat on the chopping board and cutting off the rind by pressing downwards on the knife.

⩚ To eliminate excess water from some types of squash, particularly the Piena di Napoli and Ungharese varieties (*Cucurbita moschata*), and in general those of elongated shape, cut the squash into slices and leave them to soak in salted water for about two hours, then dry them well.

Steaming

Steaming is preferable to boiling, particularly if the cooking water cannot conveniently be re-used. Cut the squash into evenly sized pieces and place in a steamer basket, making sure that they do not touch the water, when the water is already boiling. The lid should be close-fitting.

In the oven

If the squash is not too large, it may be cooked whole. To cook a squash whole, remove the stalk and put the squash to bake upside down. A sheet of baking foil laid on the bottom of the oven will collect the liquid that will drip from the squash. Cooked in this way, the flesh remains dry. It is also possible to cook squash in the oven in slices, covered with baking foil so that the surface does not dry out. Baked squash either can serve as the basis for further processes or can be eaten as soon as it is cooked, dressed with a knob of fresh butter, a drizzle of oil, or a pinch of salt.

STORAGE

❧ A whole squash will keep for months provided it is stored at a temperature no lower than about 45°F/ 8°C. It is unsuitable for freezing, which will cause it to rot.

❧ Sliced squash should be consumed within 3 or 4 days. It should be kept in the cooler, in the vegetable compartment of a refrigerator. It should above all be kept in the dark since vitamin A, in which the squash is rich, is destroyed by light.

❧ Once it is cooked (by steaming or baking), the flesh of the squash can be stored frozen in the icebox compartment of a refrigerator and used as the base for further recipes such as purée, sweet or savory pies, fillings, and so on. This applies particularly to squashes with firm, floury flesh.

In the microwave

The best way to cook squash in the microwave is sliced, and covered with clingfilm (not with baking foil). This is also the best cooking method if it is necessary for it not to become excessively dry.

Frying

Varieties of squash with firm, floury flesh can be cut into thin slices and fried in very hot oil, exactly like sautéed potatoes. Other types of squash, however, must be well dried and floured or dipped in batter, otherwise they are likely to disintegrate. The firmer types can also be fried equally well when they are floured and battered.

Stewing

Long, slow cooking in a lidded saucepan tends to result in the squash disintegrating. But this is an effective method if the squash is to be used for making a sauce to dress pasta, beans, rice, or other cooked cereal or pulses.

On the griddle

Cut the squash into evenly sized slices and cook them rapidly, first on one side then on the other. Serve them with a dressing of oil, salt and a drop of aromatic vinegar.

Some of the most suitable herbs and spices to use as a flavoring for the squash are:

In most savory dishes: *pepper, nutmeg, rosemary, sage, bay*

In certain savory dishes: *marjoram, oregano, thyme, basil, chili pepper*

In sweet dishes: *cinnamon, clove, star aniseed, ginger*

Cooking with squash

FROM THE RENAISSANCE TO THE LATE 19TH CENTURY

PLATINA
(BARTOLOMEO SACCHI)
**De la honesta voluptate
et valitudine**
PUBLISHED IN VENICE IN 1487

The squash

Mostly the squash is pot-bellied; less often it has the shape of a serpent. It grows in damp soils and hangs freely down from overhead. They say that some squashes have come to measure nine feet in length. Doctors of antiquity affirmed that the squash is congealed water: its nature is in fact cold and humid. Used as a food, the squash dilates the stomach, quenches the thirst, harms the intestine. Those of elongated form are less harmful. The peasants hollow out the big ones and dry them by smoking, then they use them for storing the seeds of vegetable. Furthermore, when the squashes are still tender they skin them, cut them into snake-shaped slices and dry them for use as food during the winter.

Fried squash

Skin the squash then cut it crosswise into thin slices. Boil them for a moment, then remove them from the pan, place them on a chopping board and leave them to dry a little. Coat them in white flour and salt and fry them in oil. Then put them on plates and pour over them a sauce made of garlic, fennel and grated breadcrumb left to marinade in unripe grape juice in such measure that the sauce is more liquid than dense. It is good to pass it through a sieve. There are those who prepare it only with unripe grape juice and fennel flowers. If you prefer it yellow, add saffron.

Squash soup

Cut a squash into slices and cook it in broth or water with a little onion. Then strain it through a perforated ladle and boil it in a pot containing fatty broth, a little unripe grape juice and saffron. When it has boiled a little, remove it from the fire, let it cool and add two yolks of egg beaten together with a little ripe cheese, grated, continuing to stir with a spoon so that lumps do not spoil the cooking. Pour it into bowls and sprinkle powdered spices on top.

Squash in milk

Boil the squash, then drain it well through a sieve or a perforated ladle and boil it in a pan containing almond milk or goat's milk. Add unripe grape juice too, or sugar, according to the taste of the company.

Squash Catalan style

Wash the squash well, then put it over the fire in a pan in which you have placed some lard. Leave it on the coals for about four hours, stirring it often with a spoon while it is boiling. Into the same vessel add fatty broth colored with saffron and seasoned with sugar and spices. There be some who also add, opportunely, two eggs beaten together with unripe grape juice and grated cheese, as I have said further up concerning the squash.

Squash cake

Wash the squash well and grate it as you do with cheese, then put it to boil a little in fatty broth or milk. When it is half cooked, pass it through a sieve and add the same quantity of cheese indicated in the preceding recipes (1 lb 8 oz/675 g), half a pound of pig's abdomen or very fat udder, boiled and pounded with a knife, or instead of these things, if haply preferred, the same quantity of lard or butter, half a pound of sugar, a little ginger, a pinch of saffron, six eggs, two cups of milk. Mix all well and tip it into an earthenware casserole lined with puff pastry and cook on a slow fire, above and below. There be some who put pieces of pastry on top and call them lasagne. When cooking is finished transfer it to a plate and sprinkle sugar and rosewater over it. Cassio, who suffers from gallstones and colic, is wary of eating this cake. In fact it is heavy to digest and is not nourishing. ◆◆

BARTOLOMEO SCAPPI
private cook to Pope Pius V,
FROM A WORK PUBLISHED IN
VENICE IN 1570

To make soup of squashes and onions in various ways ...
Take the squash and the onion and simmer them well in water until the onion, which is much harder, shall be well cooked, and take them from the water, and place them in a sieve, and let them drain, and take some grated Parmesan cheese, and fresh mozarella, and pound everything in a mortar, and dissolve it in cold broth and pass it through a sieve, and place it in a casserole, or saucepan where there is melted lard with the squashes and the onions,

letting everything boil very gently on the coals, and breaking the said squashes and onions with a wooden spoon, and stirring continuously, with the addition of a spoonful of very fatty broth or fresh butter, and fine sugar, and when it is cooked so that it is rather more solid than liquid, serve it hot with sugar on top, and sprinkled with rosewater. It can also be done in another way, that is, after they are more than half cooked in meat broth, if it be ham or salted pork belly, lift them out and pound them with a knife on a board, that is not walnut, and fry them lightly in the frying pan with lard or melted bacon fat, breaking them with a wooden spoon, and when they shall be lightly fried place them in a casserole with fatty broth, and then having beaten some yolks of egg with the juice of unripe white grapes, and mixed them with pepper, cinnamon, sugar, and saffron, and placed everything together in the casserole, and raised the mixture to boiling point stirring with the spoon, it shall be served with sugar and cinnamon on top. It can also be done in another way, that is, after they are well fried, let them finish cooking in enough goat's or cow's milk to cover them, and with sugar.

To make various soups with Turkish squashes
Pick the Turkish squash in its season, which begins in the month of October and lasts throughout April, and when it be cleaned of its rind and innards cut it in pieces and let it boil, and when it be boiled pound it with knives, and let it cook in good meat broth, and mix it with grated cheese, and beaten eggs, and it can also go well with onions in the way that our above-mentioned local custom allows. Being aware that if the squash is sound, it will be much better, and to store it it must be kept in a dry and airy place, and must not have been skinned in any place, because air will make it putrify, in this way one can make dried rind from Savona squashes after they have been boiled in hot water, and soaked in cold water.

To boil and bake the above-described squashes whole, filled with scraps.
Wishing to boil the aforesaid squashes, clean them diligently of their rinds, having a care not to break them, and make a round hole in the position of the flower or the stalk, and keep the roundel that has been taken out, and with hooked tools neatly hollow out the interior,

and when it is empty fill it with a mixture made of lean veal meat, or of pounded pork with an equal quantity of fat, and ham, and add cheese, and egg yolks, raisins, common spices and saffron, and having chickens or small pigeons without bones filled and placed in the squash with the mixture, and when it is full, stop the hole and place the squash in a pot of such a size that it cannot move, with enough broth so that it is more than half covered, with sliced ham, or salted pork belly. This is done so that the squash will take up the flavor, and not be insipid, and with the broth put in pepper, cinnamon and saffron, and cook it on the embers, keeping the pot lidded so that it cannot breathe, and when it shall have boiled for a while, until the mixture shall have melded, add more broth, letting it finish cooking, and when it is cooked, pour the broth into its own pan, and deftly move the squash onto a large plate, serving it hot with the ham or belly alongside. The said squash could even be filled with milk, beaten eggs, sugar and bruised ham cut into cubes. It could also be done in a different way; that is, having made the hole without stripping it of its rind, hollow out the interior and with dexterity surround the inside of the squash bottom and sides with bruised ham and taking chopped raw yellow sausage, or else some of the mixture, make a base at the bottom, and taking small pigeons, chickens and quails and other small birds with their insides removed and their bones crushed, and sprinkled with pepper, cinnamon, cloves and nutmeg, place them one by one in the squash, packing them in with the same sausage mixture, and finally above the said birds place a slice of veal sprinkled with the aforesaid spices, and this slice should cover the whole mixture, then stop the hole with that part of squash that was cut out, and wrap the squash with sheets of paper and tie the opening with string and place it in an oven less hot than if one were going to cook bread, and place it in such a way that it has the warm temperature all over, and let there be a base of copper underneath, or earthenware without sides. This is done so that when it is cooked the squash can be taken out without being broken, and when it has been in the oven for two hours, more or less according to size, take it out and loosen the paper, and stop the hole, and place over it another lid of raw rind, to which you have

attached a few sprigs, and serve it hot. In this squash a flavored oil can be made. ◆◆

CRISTOFORO DI MESSIBUGO
A New Book Giving Instructions for Making Every Sort of Food
PUBLISHED IN VENICE IN 1571

To make a cake of fresh squashes
Take the squashes ascertaining that they be not bitter, and clean them and grate them as you would cheese, and then put them to boil in good fatty broth with a pound of oxbone marrow or beef fat, but it need not to boil too much. Drain them through a cloth and place them in a pot with a pound of hard cheese, grated, and two ricotta cheeses, and six eggs, and a beaker of milk, and half a pound of sugar, and half an ounce of cinnamon, and a quarter of pepper, and of ginger half a quarter, and a little saffron. Mix all things together well. Then make your cake, and when it is done place on top ½ cup (120 g) of butter and put it to cook; and when it is nearly cooked, put on it ½ cup (90/120 g) of sugar and then finish the cooking. And on days that are not meat days, cook the squashes in water with butter, or in milk; and instead of the bone marrow or fat put butter. Be

it known that all the aforesaid cakes are honestly suitable for any great Prince, and for guests, and others; but for everyday, they can be made with little more than half the spices, and they will be judged good.

To make compotes of melon rinds or squash rinds, or turnips, or else peaches whole and unripe, to keep for Lent
Take a quantity of the aforesaid things according to thy mind, and clean them and place them to soak in vinegar and salt for fifteen or twenty days. Then take them from the vinegar and place them in a pot of water, and give them a good boiling; then take them out and throw them into another pot of cold water and leave them until they be cooled off. Then take them from the said water, and spread them out on a board, placing then another board on top with weights that press down; and leave them thus under pressure for a day. Then take a pot and place them in it with enough must to cover the aforesaid things, and give them a good boiling in the said must, but in such a way that they do not fall apart. Arrange them in a pot and pour over them the said must in which they have boiled; and leave them there for twenty days. Then

take another vessel and take them out of the other pot and place them in it with honey and must and cinnamon, pepper, ginger and saffron, according to the quantity that you would do, and tips of rosemary and sage; and let them boil again a little more with the aforesaid things.

Then turn them into the aforesaid pot with the said flavorings, and they will be done. And if the expense is not too heavy, this last time, in honey alone, remember that peaches should not be pressed. ◆◆

BALDASSARE PISANELLI
Treatise on the Nature of Food and Drink
PUBLISHED IN CARMAGNOLA IN 1589

Of the Squash. Ch. LXXXXVIII
It should be fresh, tender, green, and light, and sweet. It does good to cholerics, quenching the thirst if it is cooked in unripe grape juice, and it refreshes the liver. It does harm to colds, and phlegmatics, generates much windiness, and acqueousness, and weakens the stomach. Prepare with mustard, pepper and vinegar, or else with strong herbs, as are onions and parsley.

It is cold, and humid in the second degree.

It is good in very hot weather for the young, and for those who are of hot complexion.

Natural histories

The squash makes bad food, and it corrupts itself, because it changes into that humor that it finds in the stomach, and because it goes down late, and because it is insipid; it receives that flavor, and generates the humor similar to that thing, with which it is seasoned. It is bad for the intestines, especially for the colon, and therefore causes choleric pains; if eaten in much quantity it causes contractions; if eaten raw it troubles the stomach; in such a way that one cannot help oneself except by vomiting. Boiled it is much better. One type of squash can be kept all winter, but it is the food of low people. Poets call it grave, green, round, fat-bellied, pregnant, and seedy. The Latins call it *cucurbita*, and the Spanish *calabazza*.

Note LXXXXVIII

Certain reliable writers state that the use of the squash is not so pernicious as that of water melons provided that their acqueousness is corrected with appropriate things; and doctors hold that there be no better thing to mitigate the ardor of burning fevers, thirst, and relax the body, than to use often some juice of squashes cooked without water in a new earthenware cauldron, placed in the oven, at high temperatures if the squash fills up with sugar. It is a proven fact that, if the rind of the dry squash is singed and finely pulverized, putrid ulcers of the male member will be cured, if it is placed thereon. The leaves placed over the breasts of women expel the milk, and their sterile flower drives flies away from the beasts, so that they do not irritate them. Wine, placed in a hollowed out squash, and kept therein all one night, in the open, and then drunk, soothes the body. ◆◆

FRUGOLI ANTONIO
The Practice of Carving
PUBLISHED IN ROME IN 1638

Squashes, and their quality, and Cooking.
Squashes are frigid, and humid in the second degree, and the best are the long ones, they benefit cholerics, and refresh the liver, and quench the thirst, and move the body, and make defences against the phlegmatics, and to those who suffer colic pains, and contractions, because of their windiness, and harm the intestine, they will do better, and much more healthy eaten fried than boiled, because they will be considerably less humid, and watery, and the boiled ones shall be served with aromatic herbs, and cooked in good meat broth, mixed with eggs, or else almond milk, or some pine kernels, and served in divers ways, a good part of which will be said below. Squashes can be served in all ways, and dishes that are made with the melon, as I have said above in its chapter, and they shall be served like those, and they can be candied, as they do in Genoa, and they can be floured, and fried after being cut into thin slices, and pressed, and extricated from their wateriness, and served hot with bitter orange sauce, or unripe grape juice on top, or else with garlic, or almond flavoring, diluted with unripe grape juice, and in all the above ways the spaghetti of the Genoese Cocuzza can be served, after they have been soaked long enough in lukewarm water, the which will be good for covering divers boiled peeled tomatoes, after being cooked in good broth, with cernellato, or other sausages, with morsels of cheese, and unsalted Genoa mushrooms within, served with grated cheese, and spices on top, as I have said in their places. And the tender, long zucchini can be done stuffed with meat Picadilly, with pine kernels, and raisins, and unripe grapes with eggs and good young herbs beaten finely within, with sufficient spices, or else with fresh cheese pounded with good herbs, with eggs, and Parmesan cheese grated in, with spices and sugar to sufficiency, and they shall be served with grated cheese, and cinnamon on top, as a cover, and again in soup. ◆◆

BARTOLOMEO STEFANI
The Art of Good Cooking
PUBLISHED IN MANTUA IN 1662

Soup of squash tops, milk cheeses and unripe grapes together
Pick the squash tops, and if there are any zucchini there it will be better, do them in broth, and when done place them in a cauldron with capon broth, two milk cheeses cut into morsels and first done; take some of those unripe grapes that the vine usually makes three times a year, because the grapes are large and hard and have flesh, and having peeled off their skin split them, and remove the seeds, putting in ½ cup (50 g) of parmesan cheese and two eggs and thus compose the soup.

Squash soup

Take the squash done in broth, so that it shall be more flavorsome, passed through a sieve; take 1½ cups (180 g) of almonds pounded in the mortar, dissolve them with a beaker of milk poured through a sieve, placing the squash on the fire with fatty capon broth, and when the squash shall be nearly cooked, add four egg yolks and the juice of four oranges and it will taste good. ◆◆

ANTONIO NEBBIA
The Cook from Macerata
PUBLISHED IN VENICE IN 1783

Squash in various manners

Take the squash, clean it of seeds and rind, cut it into dice; make a *soffritto* of fresh herbs, marjoram and basil, mince everything finely, take a casserole, place in it 8 tablespoons (50 g) of butter, salt and sweet spices, when it is almost cooked, take a plate that you must send to table; grease the base of the plate with butter and grated parmesan cheese; proceed arranging a layer of squash and a layer of cheese, in symmetry; when finished, beat up an egg with salt and baste all the squash, and neaten the basting cleanly on top; clean the plate, place it in the country stove with the fire above, so that it makes the *brûlé*, send it to table hot with the *coulis* on top; this dish serves as an *entrée*.

Squash with polenta

Boil the squash, as you have hitherto done with salt, and when it shall be drained, take a cooking pot and put it in this with two ounces of butter, basil, and two powdered cloves, letting it fry, then pass it into a dry sieve; when it is drained, grease the base and sides of a cooking pot, take a sheet of white paper, cut it to fit the base of the cooking pot, grease it on both sides, then place it on the said base, sprinkle the inside of the cooking pot with grated bread, then take a pound and a half of sieved sugar, ten egg yolks, 1½ cups (180 g) of parmesan cheese and ¼ cup (60 g) of butter, and mix all together, then throw it into the cooking pot, place the said pot with glowing cinders below, around and above, with a lid with the aforesaid cinders; when it shall be finished, turn it out upside down into a dish, send it to table hot and tidy, that it may be enjoyed.

Squash Piedmont style

Take the squash, clean it and dice it; put chopped basil into a casserole with butter and onion; when it is lightly fried, add the diced squash with salt and powdered cinnamon, leaving it to cook, but not disintegrate; afterwards take another casserole with 12 tablespoons (90 g) of butter, to which you will add a pinch of flour, and brown it on the stove to the color of cinnamon, always stirring with a wooden spoon, add a casserole of good starch, so that it becomes a fairly dense sauce; take it off the fire, and grease the base of the dish that you will send to table; sprinkle the first layer with grated parmesan cheese, and make a layer of the said diced squash, and put on the sauce with a ladle, and so on until you have used up the squash, then take two eggs, beat them well, and with a feather brush baste the top and sides of the dish, place it in the oven so that it forms a crust; it should be sent to table hot, with the edges of the dish well cleaned.　　◆◆

VINCENZO CORRADO
The Gallant Cook
PUBLISHED IN NAPLES IN 1786

Of spring squashes

Spring squashes are of yellow color and hard rind. Summer ones can be used while they are tender, but the winter ones are better.

With citron sauce

When the squashes are cleaned of their rind and seeds, they are cut into thin slices, which being sprinkled with salt shall be purged of their malignity, as is said by others. Purged, they are floured, and fried in lard, or oil, and served with a sauce of candied citron, crushed and dissolved in lemon juice.

In fritters

The squashes are cut into pieces and cooked with butter and spices. Then they are mashed and mixed with ricotta, *cacio* cheese, eggs, spices, and a little sugar. They are shaped into nuggets, which are floured, brushed with egg, fried and served hot.

Creamed

When the squash has been cooked in butter and spices, it is mashed and sieved. Then it is mixed with milk, egg yolks, cinnamon, and nutmeg, reduced to the thickness of cream and served on crusts of bread.

In pudding

It can also be made into a pudding if first cooked with butter and spices, then mashed and sieved, then mixed with ricotta, cream of milk, egg yolks, milk, powdered cinnamon and a few breadcrumbs, and then well thickened, and once thickened it is poured into another casserole greased with butter and sprinkled with breadcrumbs, put into the oven to finish cooking, and served hot.

Of squash flowers

Squash flowers can also be served fried, and stuffed. They are fried either floured and brushed with egg, or in batter. They can be stuffed in all the ways described when speaking of the long squashes. ◆◆

ANONYMOUS
The Piedmont Cook Reduced to the Ultimate Taste and Perfection
PUBLISHED IN MILAN IN 1805

Of the squash

The squash is served in divers ways; but I will not give the description, because it is not to everyone's taste. I shall say only how to make milk soup: cook it in water; when it is cooked, and there is still a little water remaining, put in some milk, a piece of butter, salt, and sugar, if you like it, immerse bread into it, but do not let it boil. If then you wish to fry some, when the squash is cooked in the water, place it in a pan with a piece of butter, spring onion, salt

and pepper; when it has boiled for a quarter of an hour and there is no liquid remaining, add a mixture of three egg yolks with a little cream or milk.

Squashes with parmesan

Clean the squash and slice it into cubes, boil it with salt etc., drain it, and then take a casserole with ¼ cup (50 g) of butter, put the squash cubes in, with salt and sweet spices, let them fry lightly, and turn them over, and when they are mixed, place them in a dish and dress them like macaroni, that is with parmesan and butter, put them in the country stove with the fire below and above, and when they are browned, serve them.

Squash cooked in the stove

Clean and boil the squash, and drain it, and pass it through a sieve, place it in a casserole, add a little salt, with ¼ cup (50 g) of butter, ¾ cup (90 g) of parmesan, ⅓ cup (50 g) of chopped candied peel, and powdered cinnamon, let it boil, then add six beaten eggs; when it be well mixed and thickened, place it in a dish greased with butter, then brush it with two beaten eggs, brown it, sprinkle it with grated breadcrumbs, sugar and cinnamon, place it in the

country stove with the fire above, let it form a crust on top, then send it to table.   ◆◆

ANONYMOUS
Oniatology
PUBLISHED IN FLORENCE IN 1806

Cream of squash alla provençale

The squash must be yellow, and of good quality; remove its rind and the pith with the seeds, and cut it into small pieces, and cook it in water with some salt. Then drain it well and place it in a cloth, pressing it hard from both sides so that all the water comes out.

Pass it then through a fine sieve, and place what emerges in a casserole with some finely sieved flour and eight to ten egg yolks. Take two pounds (900 g) of milk, and let it boil for five minutes; when it has cooled, take out the cinnamon stick, mix it with the sugar, and put it back to cook again in the same casserole, scraping it from the sides and bottom with a ladle, until it starts to boil. Taste it and correct it as necessary; empty it into a bowl or dish and serve it at table cold. ◆◆

FRANCESCO LEONARDI
Modern Apex
PUBLISHED IN ROME IN 1807

White squash purée

Slice up a sufficient quantity of white squash, remove the rind and seeds and cut into small pieces, and place it on the fire in a pot so that it loses its water; next tip it into a sieve and drain it well, then place it in a small pot with a broth such as beans water, strongly colored so that it will give it a blonde color, a piece of ham, and an onion, with two blanched cloves, let it cook for an hour over a slow fire, skim off the fat, take out the ham and the onion, pass it through a muslin sieve, purify it near to the fire, and if before serving you wish to thicken it with the yolks of a few fresh eggs and grated parmesan, this will give it the best flavor; or else before sieving give it body with some well soaked crust of bread and a little grated parmesan.

Yellow squash purée

This is made in the same way as the foregoing, with the exception that, instead of cooking it in strongly colored broth, you use white, fairly substantial broth, then finish it as the other.

Fried yellow squash

Cut some peeled and seeded squash into strips or sections or some other way, dip it into frying batter, fry it in lard of a fine color, and serve it glazed with sugar with a red hot spatula.

Fried white squash

Clean and cut the white squash as above, season it with a little salt, then sprinkle it, flour it, and fry it to a good color, and serve it glazed with sugar with a red hot spatula. The white squash can be used in the same way as the yellow one, but the latter is preferable to the former for its flesh and color.   ◆◆

VINCENZO AGNOLETTI
Manual for the Cook and the Pastrycook
PUBLISHED IN PESARO IN 1834

White or yellow squash

Clean the squash, cut it into slices, and heat it on the fire in a casserole with butter, fine herbs, salt and spices; then sprinkle a little flour on it, and wet it with a sufficient quantity of milk; cook it over a slow fire, so that it becomes like a good thick cream, and let it cool; then add some eggs with the whites beaten until stiff, and a handful of

parmesan; tip the mixture onto the dish with a border around it, sprinkle parmesan over it, and scatter some butter, cook in a moderate oven, and serve immediately.   ◆◆

ANONYMOUS
The Unpretentious Cook or Easy and Economical Cooking
PUBLISHED IN COMO IN 1834

Squash broth

Having cooked the squash with butter, spices and salt in proportion, and afterwards drained it with a sieve, mix into it some bread crumb soaked in milk, or cream, powdered cinnamon, with the addition of a few bitter almonds, spiced cake, and a little crumbed bread. When the mixture is well thickened on the fire, pour it into the prepared mold, and bake crisp in an earthenware pot in the stove with fire above and below. Serve hot.   ◆◆

ANONYMOUS
The Modern Italian Cook
PUBLISHED IN LIVORNO IN 844

Squash soup

Take some white squash, clean it and cut it in slices half a finger thick, then into tiny dice, put these into a

casserole with a good piece of butter, letting it fry over a low fire until it has taken on a fine golden color. Then chop some onion very fine with parsley, basil, celery, thyme, a little garlic, and put it all into the squash, stirring well, and adding two cloves, water as needed, fish juice or, failing that, butter or oil, or the one and the other together. Let it boil thus for the space of an hour, then scatter on top some fingers of toast or fried bread done in butter or oil according to taste.

◆◆

ANONYMOUS
The Cookery Book for Town and Country or New Economical Cooking
PUBLISHED IN TURIN IN 1845

Various species of squash with parmesan. (Citrouille et potiron á la parmesane). (Handed down).

Cut it into square pieces, boil it for quarter of an hour in salted water; take it out and let it drain. Put a good knob of butter in a casserole, and there fry your pieces with salt and spices; take them out and arrange them on a plate and cover them with grated cheese. Let them take color under a lid with fire above and serve.

Various squashes cooked in the oven. (Citrouille ou potiron au four). (Handed down).

Cook them in salted water then juice them. Put this in a pan with ¼ cup (50 g) of butter, 3 ounces (90 g) of cheese, ¼ cup (50 g) of sugar and some powdered cinnamon. Bring to the boil; add six beaten eggs; mix all together and place on a dish greased with butter. Brush the top with egg, sprinkle with breadcrumbs mixed with sugar and cinnamon, and let it brown in the oven or under the grill.

Soup of lumpy squash reduced to sauce. (Potiron en purée)

Cut the squash into pieces and place it in boiling water for five minutes with salt. Take it off the fire, discard the water; then press the squash with a rolling pin or similar, then melt some butter in a casserole and place the squash in it to scorch a little. In the soup tureen place slices of toasted bread spread with butter and sugared; then pour boiling milk over them, adding the squash. Mix all together and serve it at table, after having let it warm over a slow fire, if you so please.   ◆◆

G.F. LURASCHI
New Economical Cook of Milan
PUBLISHED IN MILAN IN 1853

Trifoliate big-cap squashes
Clean the squashes, cut them up and blanch them in salt water and drain them in a sieve. Lightly fry a bunch of parsley, a clove of garlic, a little shallot and three cloves, all chopped fine, add the squashes and put in a little pepper, nutmeg and a little sauce, leave them a while to simmer, skim off the fat and serve with croûtons of bread toasted on the gridiron. ◆◆

ANONYMOUS
Book of 14th-century Recipes edited by F. Zambrini.
PUBLISHED IN BOLOGNA IN 1863

About the squashes
Take new squashes cut and washed in hot water, and press them strongly in a cloth, and put them to cook with fresh pig's meat, and pepper and saffron.

Alternatively. Again take new squashes, and wash them and press them strongly, and with cooked eggs, and with onions, and *cascio* cheese well minced, and plunge them into boiling water, with

pepper and with saffron, and sufficient garlic, and some salt. And with such one can make ravioli with mixed tenderized meat, and also pasties.

Alternatively. Take dried squashes, and put them to soak in hot water, in the evening; and when they are softened, cut them minutely, and cut on the board, with onions, and with garlic, pepper and saffron; fry lightly and put them to cook in civet, made of vinegar and bread crumb. And one can do it in this way with almond milk, pepper, saffron, salt and garlic and with walnut milk. ◆◆

VIALARDI GIOVANNI
Simple and Economic Bourgeois Cooking
PUBLISHED IN TURIN IN 1863

Squash alla contadina (country style).
Take a fine squash, green and tender, cut it into long slices, remove the innards and the green skin, cut it into very fine slices, fry it in a pan over a hot fire in 1 hectogramme (½ cup/120 g) of butter, a little chopped garlic, salt, pepper and spices dampened with a little water if necessary until it is cooked tender

and pale; mix in some cheese and a little vinegar and serve.

Squash cake alla giardiniera (as made by the gardener's wife)
Take 8 hectograms (3½ cups/810 g) of tender Roman squash, remove the innards and the hard yellow rind, cut it into small slices or grate it, boil it in water for five minutes, or until it is tender; drain it, put it in a pan with 2 hectograms (1 cup/200 g) of butter, a little garlic and chopped parsley; fried, dried, and slightly browned, add ¼ cup (60 grams) of flour and fry a little more, pour over it 2 cups (half a liter) of sour milk and let it cook until reduced and thick; remove from the fire, crush six amaretti and mix them in, plus a little cascio cheese, salt, pepper, spices and six whole eggs, all being well mixed, tip it into an earthenware dish, greased with butter and fireproof, let it cook slowly with the fire above and below, or better in the oven until the cake is firm in the middle, risen by a quarter and of a fine golden color, and serve it on the same dish. Instead of the earthenware dish it can be put in a mold greased with butter, sprinkled well with bread, cooked in the oven and turned out onto a plate. ◆◆

ANONYMOUS
The Mistress Cook
PUBLISHED IN REGGIO EMILIA IN 1886

Squash cake
Take some squash, peel it and scrape it then put it in a linen cloth to drain of its own water, and then take ½ cup (120 g) of butter and ½ cup (120 g) of dripping and make a mixture, and lightly fry it with a little onion, a little pepper or spices in a casserole then put in the squash and let it boil for quarter of an hour then take it off the fire and leave it to cool in a pan and when it is cool add a pound (450 g) of ricotta. Then take a pound (450 g) of almonds, skin them and crush them in a mortar and unite them in the pan with 1 cup (225 g) of sugar, eight eggs, four egg yolks: then beat all the things together very well, then make your sweet pastry, lay it out in the cake tin greased with butter, empty the aforesaid mixture into it and cook it in the oven.

Squash with parmesan
Clean the squash and cut it into small cubes, simmer it with salt, then drain it; then take a casserole, put a piece of butter in it, the cubed squash with salt and spices, fry them

lightly and turn them over; when they are blended, place them in a dish and dress them with parmesan and butter, place them in the oven with fire above and below; and when they have taken on a good color, serve. ◆◆

Jean Marie Parmentier
The King of Kings of Cooks
Published in Milan in 1897

Squash soup (potage au potiron)
Take a quart (2 pt/1 l) of squash, remove the rind and the seeds; cut it into walnut-sized pieces and place it on the fire in a stock-pot with some water. When the squash is well reduced to a purée, add ¼ cup (62 grams) of butter and a little salt. Turn up the heat a little.

Bring 4 cups (1 litre) of milk to the boil, add a little sugar or salt, if you prefer it, and stir it into the squash purée. Cut some bread into small pieces and put them in the tureen and pour over them the mixture of squash and milk.

◆◆

Pellegrino Artusi
Science in the Kitchen and the Art of Eating Well
Published in Florence in 1899

Yellow squash cake
This cake is made in autumn or winter, when the yellow squash is found on sale at market gardens.
  Squash, 2 lb 3 oz (1 kilogram)
  Sweet almonds, 3½ oz (100 grams)
  Sugar, ½ cup (100 grams)
  Butter, ¼ cup (30 grams)
  Breadcrumbs, 30 grams (1 oz)
  Milk, 2 cups (half a liter)
  Eggs, 3
  A pinch of salt
  Powdered cinnamon flavoring.

Peel the squash, clean off the superficial fibers and grate the flesh onto a linen cloth. Take the four corners of this to collect it together and wring it so as to remove a good part of the moisture that it contains. The squash will then be reduced to about a third. Then put it to boil in the milk until it is cooked, which may take between 25 and 40 minutes, depending on the quality of the squash. Meanwhile, having skinned the almonds, pound them, together with the sugar, in a mortar, reducing them to a very fine consistency, and when the squash is cooked combine all the ingredients except the eggs, which you will add when the mixture is chilled. Grease a pie dish generously with dripping and line it with a sheet of shortcrust pastry and onto this pour the mixture to a depth of about one and a half fingers, cooking it between two fires or in the oven. I recommend a moderate heat and the precaution of a sheet of paper over it, greased with butter.

Yellow squash soup
A yellow squash, skinned and thinly sliced, 2 lb 3 oz (one kilogram). Put it to cook with two ladles of broth and then drain it in a sieve. On the fire make a paste with ¼ cup (60 grams) of butter and two level spoonfuls of flour, and when it has turned golden quench it with broth; add the drained squash and the rest of the broth, which will be enough for six persons. Then pour it boiling over fried bread cubes and send the soup to table with grated parmesan on the side. If you make this soup properly and with good broth, it can appear on any table and it will also have the virtue of being refreshing. ◆◆

Recipes by Arneo Nizzoli

Arneo Nizzoli is the acknowledged King of the Squash, a title he has won through his expertise with this vegetable. His kingdom is at Villastrada, a district of Dosolo, between Viadana and Mantua. Here flows the River Po, its banks lined with rows of poplars, and a little further away is the confluence with the Oglio. This is one of the most characteristic corners of Bassa Padana (the Lower Lombardy Plain), strongly marked by the presence of the river.

On either side of the Po are lands that evoke many episodes in Italian history, from the luxurious excesses of the Gonzaga to the battles of the Risorgimento, from the peasants' struggles in the early 20th century to the heroism of the Cervi brothers, and the romantic rivalry between Don Camillo and Peppone.

Through Nizzoli we can conjure up the atmosphere of the Lombardy Plain, and enjoy the foods of that locality, from the rich dishes that suggest a taste of the Renaissance (specifically the taste of the Gonzaga nobility) to the humble yet ingenious dishes of the peasant tradition.

For rich and poor alike, the squash holds a place of honor. But to savor it in full go to Dosolo in October, which Arneo Nizzoli proclaims as the month of the squash. Meanwhile enjoy a foretaste by reading his recipes and trying them out.

NB: *Some of the squashes illustrated on the following pages are purely ornamental.*

Stuffed squash flowers

5 potatoes, 7 oz/200 g of beans,
7 oz/200 g of zucchini flowers,
2 eggs, 1 clove of garlic,
1 tbsp of parmesan (to taste),
1 tbsp of oil, salt.

SERVES 6

WINES – **For this dish choose a dry white wine with a slightly aromatic bouquet, such as Vermentino from the western Ligurian Riviera, Bianco Vergine Valdichiana, Trebbiano d'Abbruzzo or Contessa Entellina Chardonnay.**

METHOD

1. Boil the vegetables, strain and mash them to produce a purée. Add the eggs, garlic, chopped parsley, salt and parmesan to taste.
2. Wash the flowers well and fill them with the prepared mixture, arrange them in a baking dish and cook them in the oven at 355°F /180°F for about 40 minutes.

PER PORTION:
125 calories, fiber 0.1 oz/2.9 g

Squash rinds

COOKED IN APPLE MOSTARDA

The rind of 1 squash, 2 pinches of salt, 1 tbsp of sugar,
1 stick of cinnamon, 1 tbsp of vinegar,
3 tbsp of apple mostarda.

SERVES 6

WINES - **This stimulating dish, which combines sweet, sour and spicy flavors, can be accompanied by a young dry white wine, sweet or medium sweet, with an intense bouquet, such as the aromatic late harvest Alto Adige Traminer, the sweet Pagadebit from Romagna or the medium sweet Orvieto.**

METHOD

1. Wash the whole squash, open it, and cut it into walnut-sized pieces. Scald for 5 minutes in boiling water.

2. Cut away the rind and place it in an infusion of salt, sugar, cinnamon, white wine vinegar and three tablespoons of Mantuan apple mostarda.

3. Leave to macerate for about 2 hours. Serve as a starter, with a sprinkling of parmesan if desired.

PER PORTION*

* The nutritional value of the squash rind cannot be calculated but it should depend on its fiber content rather than its calorific value.

Fried halfmoons of squash

2 lb 3 oz/1 kg of squash, rind and seeds removed,
flour, salt, garlic, rosemary,
a cup of oil for frying

Serves 6

WINES - A suitable wine for this extremely simple dish will be a young dry white with an intense, fruity bouquet, such as the Langhe Favorita, the Custoza Blanco, or the white Epomeo.

METHOD

1. Cut the squash into slices ½ inch/1 cm thick, shaped like halfmoons. Lay them in a dish, sprinkle them with salt and leave them for 1 hour.
2. Wash, dry and flour the halfmoons, then fry them in plenty of oil flavored with garlic and rosemary.
3. Drain the halfmoons on a sheet of kitchen paper, season with salt and serve hot.

PER PORTION:
110 calories, fiber 0.05 oz/1.4 g

Squash in hot sauce

2 lb 3 oz/1 kg of squash, rind and seeds removed,
2 cloves of garlic, sage, rosemary, 1 cup of vinegar,
2 tbsp of breadcrumbs, flour, oil for frying,
salt, pepper.

SERVES 6

WINES - **The gently pungent flavor of the sauce, the strong aroma of the ingredients, and subtle sweetness of the squash make a dry white wine with a strong fruity, flowery bouquet a suitable accompaniment for this dish. Choose a soft, refreshingly sharp wine, strongly alcoholic and full bodied, such as the white Breganze Pinot, the white Montecarlo or the Bianchello del Metauro.**

METHOD

1. Cut the squash into thin slices, sprinkle with salt and leave for 1 hour.
2. Flour the slices and fry them in plenty of hot oil, then dry them on kitchen paper.
3. Chop the garlic and fry it separately in a little oil, then add the herbs and 2 tbsp of breadcrumbs. Fry for a little longer, then add the vinegar, salt and pepper. Leave to cool.
4. Arrange the slices of squash on a serving dish and dress them with the sauce.

PER PORTION:
137 calories, fiber 0.04 oz/1.1 g

Squash boats

WITH ANCHOVIES

1 lb 12 oz/1 kg of squash, rind and seeds removed,
3½ oz/100 g of anchovies, 2 tbsp of oil,
1 tbsp of balsam vinegar,
and salt, pepper, and sugar.

SERVES 6

WINES - The highly flavored, aromatic, and spicy characteristics of this dish demands a mature, dry, white wine with a strongly fruity bouquet, such as the Florano Semillion, the Falanghina del Sannio or the white Etna.

METHOD

1. Cut the squash into slices about ¾ in/1.5 cm thick.
2. Separately prepare a sauce of oil, salt, pepper, sugar and vinegar, and sprinkle it over the base of an ovenproof dish.
3. Arrange the squash boats in the dish, lay an anchovy on each one and bake in the oven at 355°F /180°C for about 15 minutes.
4. Remove from the oven and serve warm.

PER PORTION:
83 calories, fiber 0.02 oz/0.7 g

Tasty squash slices

1 lb 12 oz /1kg of squash, rind and seeds removed,
5¼ oz/150 g of ham, 7 oz/200 g of mozzarella,
2 eggs, flour, sage, 2 cups/½ l of fairly liquid béchamel,
salt, oil for frying.

SERVES 6

WINES - **The tendency towards sweetness in this dish is balanced by the succulent flavors of the mozzarella and the ham. The accompanying wine might be a young rosé with a fruity bouquet, such as the Alto Adige Lagrein rose, the Rosa Cormòns or the Rosa del Golfo.**

METHOD

1. Cut the squash into evenly sized squares. Dip them in egg and flour, then fry them.
2. Remove them from the pan, and when they have cooled, lay a slice of mozzarella, a small slice of ham, and a leaf of sage on each. Cover with another square of squash.
3. Arrange the squares in an ovenproof dish, cover them with béchamel and cook them au gratin in the oven at 355°F/180°C for about 20 minutes. Serve very hot.

PER PORTION:
435 calories, fiber 0.04 oz/1.2 g

Squash fritters

1lb/500 g of squash, the whites of 3 eggs,
1 pinch of flour, 1 cup/100 g of grated cheese,
salt, nutmeg, 1 tbsp of saffron,
oil for frying.

SERVES 6

WINES - **The ideal accompaniment to this spicy and succulent dish is a young dry white wine, with a fairly intense, fruity bouquet, such as Sandbicheler, Bolgheri or Vermentino di Alghero.**

METHOD

1.Cut the squash into four, remove the rind and seeds, and cut it into julienne strips using a vegetable slicer.

2. Lightly whisk the egg whites and fold in the flour, the grated cheese, and the salt, nutmeg, and saffron.

3. Add the squash to this mixture. Spoon it out, slipping each spoonful into the boiling oil to fry.

4. Drain on kitchen paper, sprinkle with salt and serve.

PER PORTION:

143 calories, fiber 0.02 oz/0.5 g

Squash and onion tartlets

1 lb 12 oz/800 g of squash, rind and seeds removed,
1¼ cups/300 g of tomato sauce, 4 onions,
oil, oregano, bay, salt,
6 pastry tart cases.

SERVES 6

Wines - This simple but delicious dish combines sweet, aromatic, and spicy flavors and a barely perceptible oiliness. As an accompaniment choose a young, dry white wine with an intense bouquet, such as Colli Berici Tocai, Bianco di Pitigliano or Regaleall.

METHOD

1. Dice the squash and the onion.
2. Toss them in a frying pan with plenty of oil, taking care that the squash does not disintegrate. Add the salt, oregano, and tomato sauce.
3. Arrange the mixture in the pastry cases and place them in a hot oven to brown.
4. Serve garnished with a bay leaf.

PER PORTION:
224 calories, fiber 0.14 oz/3.9 g

Savory squash pie

*2 lb 3 oz (1 kg) of squash, 9 oz (250 g) of provolone or
mozarella, 5 slices of sliced white bread, 1 egg,
ground cinnamon, salt.*

For the pasta:

*3 cups (350 g) of plain white flour, 1¼ cups (150 g) of butter,
1 egg, salt.*

SERVES 10–12

WINES - **This dish may be accompanied by a dry rosé with a fruity,
flowery bouquet, such as Chiaretto di Moniga, Lagrelnrose, Salice
Salentino or Settisoli.**

METHOD

1. Pour the flour in a heap on a pastry board, add the butter, softened and cut into cubes, and rub in quickly. Then add 3 or 4 tbsp of cold water, the egg and a pinch of salt. Work the pastry as little as possible. When all the ingredients are combined, lightly press the pastry into a ball, cover it with clingfilm and place it in the refrigerator for about 20 minutes.

2. Peel the squash and cook it in boiling salt water for about 15 minutes, then drain it and leave it to cool.

3. Grease and flour a 10 in/25 cm hinged cake tin; roll out the pastry to a thickness of about ⅛ inch/3 mm and use half of it to line the cake tin.

4. Place a layer of sliced bread on the bottom of the cake tin, add a layer of thinly sliced squash, sprinkle with salt and cinnamon, then cover with a layer of thinly sliced cheese. Repeat the process with the rest of the squash and cheese.

5. Preheat the oven to 355°F/180°C. Cover the layers of squash and cheese with the remaining pastry. Seal it well, pinching it to the dish all round the sides, and use the offcuts to decorate the top. Paint the top with egg, then place the pie in the oven and cook for about 50 minutes.

6. Serve hot or warm.

PER PORTION:
347 calories, fiber 0.06 oz/1.7 g

Onion and squash omelet

10½ oz/300 g of squash, rind and seeds removed, 1 onion,
10 eggs, 1½ cups/100g of grana or parmesan, grated, chives,
olive oil, salt, red wine vinegar.

SERVES 6

WINES - **The ingredients and the cooking method of this dish give it a pleasant, fragrant aroma that blends elegantly with the sweetness of the squash, the egg and the onion. Choose a dry white wine with an intense, fragrant bouquet, such as white Folaneghe, Piave Pinot Grigio, or Castel del Monte.**

METHOD

1. Shred the squash and the onion and fry in the oil until golden.
2. Beat the eggs and salt them.
3. Drain the squash and onion of the oil and add them to the beaten eggs. Add the grated cheese and the chopped chives.
4. Pour the mixture in an oiled frying pan large enough to allow it to spread out thinly. When cooked, dry the omelet on kitchen paper.
5. Serve hot or warm, sprinkled with a drop of red wine vinegar.

PER PORTION:
286 calories, fiber 0.035 oz/1 g

Pasta filled with squash

AND RICOTTA CHEESE

For the pasta:

2½ cups/300 g of flour, 3 eggs, 2 tbsp of extra virgin olive oil

For the filling:

8 oz/225 g of yellow squash, diced, 8 oz/225 g of sheep's-milk ricotta, ¼ tsp/8 g of Dijon mustard, ½ oz/10 g of onion, 1 tsp of thyme, 1 tbsp of oil, salt, pepper, 1 leek cut into strips lengthwise, 1 tbsp of butter,

2 tbsp of grated parmesan.

For the garnish:

8 oz/225 g of mushrooms, ⅓ cup/75 g of tomato, cubed, 1 tbsp of shallot, 1 tsp of thyme, 1 tbsp of oil.

For the sauce:

5 tbsp of meat broth, 1 tbsp of oil, saffron

SERVES 6

METHOD

1. Prepare the pasta; pull it out into a thin sheet and cut it into 8 in/ 200 cm squares. Cook them *al dente* in boiling salted water; drain and leave to cool on a board.

2. Cook the strips of leek for 1 minute in boiling salted water, drain and leave to cool.

3. Toss the squash in a frying pan with the oil and onion. Remove from the pan, leave to cool, and mix with the ricotta. Add the thyme, mustard, salt and pepper.

4. Spoon some of the filling into the center of each sheet of pasta; lift the four corners and tie into bags with the strips of leek.

5. Preheat the oven to 355°F/180°C. Arrange the pasta cases in a buttered ovenproof dish, sprinkle them with grated parmesan and place them in the oven for 20–25 minutes until brown.

6. Slice the mushrooms and toss them in a frying pan with the oil, thyme, salt, shallot, pepper and the diced tomato. Remove from the pan and keep warm in a tureen.

7. In the same pan, add the saffron to the broth, reduce, then emulsify with a little oil.

8. Place the pasta cases onto plates, garnish with the mushrooms and dress them with the sauce.

PER PORTION:

464 calories, fiber 2.8 g

WINES - **For this sweet, slightly oily dish choose a dry white wine with an aromatic bouquet such as Terlano Sauvignon, Collie Chardonnay, Biancolella d'Ischia or Nuragus di Cagliari.**

Squash tortelli

For the pasta:

a generous 4 cups/500 g of flour, 4 eggs, warm water

For the filling: *a generous 1lb/500 g of squash, 7 oz/200 g of apple mostarda, 3½ oz/100 g of macaroons, a pinch of grated nutmeg, the rind of half a lemon, grated, breadcrumbs, 1 medium onion, ½ cup/100 g of lard, ¼ cup/50 g of butter, 1 glass of dry white wine, ½ tbsp of tomato concentrate, salt and pepper.*

SERVES 6–8

WINES - **Since the squash and the pasta are sweet, the filling spicy and succulent, the accompanying wine should be white and young, with a flowery, lightly spiced bouquet, such as Valle Isarco Gewürztraminer, Valle d'Aosta Petite Arvine, Poggio alle Gazze di Bolgheri or Regaleali Nozze d'Oro.**

METHOD

1. To prepare the filling, cook the squash and reduce it to a pulp. (If boiled, leave it to drain in a colander for 2 hours.) Add the mostarda, finely chopped, with some of its liquid; the macaroons, finely crushed; the lemon rind, nutmeg, salt, and pepper; and a few breadcrumbs to bind the mixture. Cover and leave to rest for about 2 hours.

2. To prepare the pasta, work the eggs into the flour with a pinch of salt and a little water. Knead energetically to obtain a smooth, elastic and firm mixture, wrap in a cloth and leave to rest for about 30 minutes.

3. Knead it again then roll it out into a thin sheet and cut it into rectangles about 4 in/100 cm long. Spoon some of the squash mixture onto each rectangle and firmly seal the long edges.

4. Prepare a fried mixture of lard and butter, chopped onion, white wine, tomato concentrate, salt and pepper. The resultant dressing should be left to reduce for at least 1 hour, with a little water added if necessary.

5. Boil the *tortelli* in plenty of salted water, remove them with a perforated ladle as soon as they are cooked. Take a warmed tureen and cover the bottom with a little of the sauce and a sprinkling of grated cheese, then a layer of *tortelli*, followed by a further layer of sauce and cheese until the tureen is full. Cover with a cloth and leave to rest for about ½ hour, then serve.

PER PORTION:

546 calories, fiber 0.15 oz/4.3 g

Squash soup

6 cups of chicken broth, 6½ lbs/3 kg of squash, rind and seeds removed, 1 small onion, 3 leeks, 3 floury potatoes, peeled, 3 tbsp of flour, 2 cups/ ½ l of single cream, salt, pepper, 3½ oz/100 g) of gruyere, 10 tbsp of butter, 2 cups of croûtons.
Serves 10–12

WINES - **The dominant characteristic of this dish is its tendency to sweetness, and a delicate flavor given by the leeks and onion. Choose a young, dry, white wine, with a fragrant bouquet, such as Lugana, Lessini Durello, Pinot Bianco di Monte San Pietro or Orvieto classico.**

METHOD

1. Clean, wash, and finely slice the onion and leeks and toss them in the butter. Add the flour and stir it in.
2. Cut up the potatoes and squash, add them to the onion and leeks, and brown for 15 minutes.
3. Season with salt and pepper, add the chicken broth, bring to a boil and leave to cook for 1 hour over a moderate heat.
4. Pass through a sieve. Return the soup to the heat, add the cream, and check the seasoning.
5. As soon as it starts to boil, remove the soup from the heat and add the grated gruyere.
6. Serve with the croûtons and clots of whipped cream (optional).

PER PORTION:

512 calories, fiber 0.15 oz/4.3 g

Spaghetti with squash

1¼ lbs/540 g of spaghetti, 10½ oz/300 g of mascarpone,
grated parmesan to taste,
a slice of squash weighing 1 lb/500 g, 2 shallots,
1 nut of butter, sage (leaf and ground), rosemary,
salt, pepper.

SERVES 6

WINES - **This pleasing dish should be accompanied by a dry white wine, fairly mature (and possibly sparkling) such as Breganze di Breganze, Cervaro della Sala, white Florano or Greco di Tufo.**

METHOD

1. Fill a large pan with plenty of salted water and bring to a boil.
2. Clean the squash and cut it into small slices. Chop the shallots and toss them lightly in a pan with a small lump of butter. Then add the slices of squash and two sage leaves; when these have imparted their flavor, remove them.
3. Cook the spaghetti in the boiling water. Put the mascarpone and the parmesan into the serving dish, and mix them with a ladle of water from the spaghetti pan. Add a generous quantity of ground sage and rosemary.
4. When the pasta is *al dente*, drain it and add it to the serving dish, with the cream. Add the squash, mix, and serve.

PER PORTION:
589 calories, fiber 1.2 oz/3.4 g

Potato soup with diced squash

1lb/450 g of white potatoes, 3½ oz/100 g of chopped onion,
10½ oz/300 g of squash, diced, 6 slices of smoked bacon,
shredded, 4 cups of vegetable broth, 10 tbsp of butter,
1 tbsp of chopped parsley.

SERVES 6

WINES - **An appropriate wine for this dish is a young dry white such as Erbaluce di Caluso, Valcalepio, Bianco Vergine Valdichiana or Montecompatri Colonna Superiore.**

METHOD

1. Peel and slice the potatoes. Lightly brown the onion in 6 tbsp of butter. Add the potatoes, season with pepper, and cover with vegetable broth.

2. When the potatoes are soft, pass the mixture through a vegetable mill and add salt and pepper.

3. Melt the rest of the butter in a frying pan, toss the bacon in it and add the squash with 4 tbsp of vegetable broth, taking care not to overcook the squash.

4. Serve the potato soup in bowls with a spoonful of diced squash and bacon and a pinch of chopped parsley in the middle.

PER PORTION:
327 calories, fiber 0.09 oz/2.7 g

Rigatoni with squash

1lb 5 oz/600 g of rigatoni, 1 lb/450 g of onion, sliced,
the heart of 1 celery, chopped, 3 lb 5 oz/1.5 kg of squash,
diced, ⅔ cup/150 g of butter, 6 ripe tomatoes,
salt, pepper, grated parmesan.

SERVES 6

WINES - Choose a young, dry, white wine, such as Colli Orientali del Friuli Riesling, Colli di Luni, Falerio dei Colli Ascolani or Falanghina del Sannio.

METHOD

1. Into a wide saucepan put ½ cup/120 g of butter, the onion, the celery and the squash and let them cook, without the lid and on a high heat, until the squash has softened. Add the peeled and chopped tomatoes, salt and pepper, and cook on a low heat.

2. Meanwhile boil the rigatoni in plenty of salted water.

3. Shortly before serving, add the remaining butter to the sauce. Drain and add the rigatoni, and sprinkle with parmesan.

PER PORTION:
626 calories, fiber 0.29 oz/8.4 g

Squash gnocchi

3 lb 5 oz/1.5 kg of squash, baked and cooled, 2¼ cups of salted water, plain white flour, sage, butter, grana or parmesan, salt.

SERVES 6

WINES - An appropriate wine for this simple dish is a fresh young white with a fairly intense, fragrant bouquet, such as Cortese di Gavi, Bianco di Custoza, Pomino or Leverano.

METHOD

1. Pass the squash through a sieve and heat in a pan with the salted water. Gradually add enough flour to make a polenta; it is cooked when it comes away from the sides of the saucepan.
2. Turn the mixture out onto a board and let it cool. Divide it and shape into gnocchi, with a little flour.
3. Cook the gnocchi in plenty of salted water and dress them with melted butter, sage, and grated cheese.

PER PORTION:

270 calories, fiber 0.06 oz/1.8 g

Squash risotto

2 cups/400 g of rice, 14 oz/400 g of squash,
meat broth (double the volume of the rice),
⅔ cup/100 g of butter, 3½ oz/100 g of grated parmesan.

Serves 6

WINES - This is a classic dish from the Lombardy Plain. It may be accompanied with a fresh young red wine, possibly sparkling, such as Grignolino d'Asti, Rosso dell'Oltrepò Pavese, Lambrusco Salamino di Santacroce or Colli del Trasimeno.

METHOD

1. Cut the squash into thick slices, remove the seeds, and cook them in a little salted water. Then drain, allow to cool, and remove the rind.
2. Put the rice, the squash, and the cold broth in a saucepan, stir well and cook for 15 minutes.
3. When the rice is cooked, add the butter and stir until it has melted. Add the parmesan and serve. The flavor can be enhanced with a spoonful of brandy, if desired.

PER PORTION:
503 calories, fiber 0.06 oz/1.8 g

Squash lasagne

A generous pound/500 g of fresh lasagne,
2 lb 3 oz/1 kg of squash, 1 medium onion, chopped,
1 tbsp of butter, 1 tbsp of olive oil,
1 glass of sweet white wine, 2 cups of béchamel,
7 oz/200 g of mortadella, diced, grated parmesan,
nutmeg, salt and pepper.
SERVES 6

WINES - **The accompanying wine should be a refreshing young red, with a fruity bouquet, such as Freisa di Chieri, Valcalepio, Lison-Pramaggiore Refosco, Sangiovese di Aprilla or Pentro di Isernia red.**

METHOD

1. Boil the squash in salted water.
2. Cook the lasagne *al dente* in boiling salted water with a drop of oil, drain and leave to dry on a cloth.
3. Meanwhile, in a saucepan gently fry the onion in the butter and olive oil, then add a glass of sweet white wine.
4. Drain the squash, reserving the cooking water, and remove the rind. Add the squash to the saucepan, breaking it up well with a whisk to eliminate all lumps. Add some of the cooking water if needed.
5. Then add the béchamel, the mortadella and season with nutmeg, salt and pepper. (The resulting mixture should be fairly dense.)
6. Cover the bottom of a baking dish with lasagne, followed by a layer of the squash mixture and parmesan. Repeat until all of the ingredients have been used, ending with a layer of lasagne.

7. Dot with butter and brown in a hot oven, about 390°F/200°C, for 30 minutes.

PER PORTION:
516 calories, fiber 0.15 oz/4.2 g

Gnocchi

WITH BEANS AND SQUASH SOUP

For the soup:

1lb 5 oz/600 g of squash, rind and seeds removed and diced,
7 oz/200 g of cooked beans, ½ an onion, finely chopped,
4 ripe tomatoes, salt & pepper, 1 tbsp butter, 1 tbsp olive oil.

For the gnocchi:

2½ cups/300 g of plain white flour, warm water, salt
SERVES 6

WINES - **A suitable accompaniment for this unusual and appetizing dish might be a young, dry red wine, possibly sparkling, such as Malvasia di Casorzo d'Asti, Oltrepò Pavese Croatina, Priuli Grave Merlot, Chianti di Montalbano or Solopaca.**

METHOD

1. Place all the ingredients for the soup in a pan with a little cold water, bring to a boil, and simmer for about 2 hours, until the mixture is well condensed.

2. Work together the flour, salt and water, keeping the mixture smooth and not sticky. Shape the pasta into rolls and then slice them, aiming to make the gnocchi about the size of hazelnuts.

3. Cook the gnocchi in salted water, drain and add to the prepared soup. Serve with a sprinkling of grated parmesan and black pepper to taste.

PER PORTION:

266 calories, fiber 0.17 oz/4.8 g

Sausage and squash morsels

ALL' ERBA CIPOLLINA

5¼ g (150 g) of sausage, 2 lb 3 oz (1 kg) of squash,
1 tbsp of butter, 1 tbsp of oil,
a bunch of fresh chives, salt, nutmeg.

SERVES 6

WINES - **For this dish, dominated by the texture of the sausage and the flavor of the chives, choose young dry red wine, full-bodied with a strong fruity nose, such as Valle d'Aosta Chambave, Santa Maddalena, Colli Berici Tokay, Montescudaio or San Severo.**

METHOD

1. Cut the sausage into bite-sized pieces and dice the squash (the pieces should not be too small). Fry them lightly in oil and butter and season with salt and nutmeg.

2. When cooked, add the chopped chives and mix together well, taking care that the squash does not disintegrate.

PER PORTION:
137 calories, fiber 0.03 oz/0.9 g

Chicken breasts with squash

AND FORTIFIED WINE

*3 chicken breasts, 3 cups (800 g) of sugar,
1 tbsp of butter, 1 tbsp of oil, 1 sprig of fresh sage,
3 tbsp of fortified wine such as port or sherry.*

SERVES 6

WINES -The subtle aroma of the sage and the delicate flavor of the chicken and the squash make this a light, pleasant and distinctive dish. Red or white wine is equally suitable as an accompaniment; choose a young, light wine such as Oltrepò Pavese Cortese, Aquillela Verduzzo or Feudo dei Fiori, or else Barbera del Monteferrato, Lago di Caldano or red Copertino.

METHOD

1. Cut the chicken into small pieces and dice the squash. Fry them in oil and butter and flavor with a few sage leaves.
2. When half cooked add 3 tbsp of fortified wine and finish cooking.
3. Serve with fried squares of polenta.

PER PORTION:
516 calories, fiber 0.15 oz/4.2 g

Braised pork with squash

12 oz/350 g of filleted pork,
10 oz/300 g of squash, rind and seeds removed and diced,
oil, butter, broth, salt, pepper,
1 glass of white wine, cane sugar or honey.

SERVES 4

WINES - **The accompanying wine for this succulent dish could be a young rosé, such as Alto Adige Legrein, Rosato Castel Grifone, Rosato di Irpinia or Settesoli.**

METHOD

1. Fry the meat over a strong heat with a little oil and butter until browned.
2. Add the squash and stir, adding a ladle of broth, then the wine and a little cane sugar or honey. Season with salt and pepper and serve very hot.

PER PORTION:
223 calories, fiber 0.02 oz/0.5 g

Vanilla cotechino

WITH SQUASH PURÉE

1 cotechino (spiced pork sausage) weighing about 2 lbs/1 kg, 2 sachets of vanilla, a generous 1lb/500g of squash cut into two slices about 2 in/4 cm thick, 7 tbsp of butter, milk, salt, nutmeg and parmesan.

SERVES 6

Wines - A suitable wine for this succulent dish with a slight tendency to oiliness is a fresh young red wine, such as Oltrepò Barbera, Gutturnio dei Colli Piacentini, Lambrusco di Sorbara or Marzemino di Isera.

METHOD

1. To cook the *cotechino*: wrap it in a cloth or a sheet of greaseproof paper, secure with string, and place in plenty of cold water with the vanilla, which will give flavor and counteract the greasiness of the *cotechino*. Bring to a boil and simmer for about 3 hours.

2. To make the purée: boil the squash in salted water and, as soon as it is cooked, drain and remove the rind. Pass it through a vegetable mill, put it in a saucepan, and add a little milk, salt and nutmeg. Cook on a low heat, stirring constantly. Just before serving, add the butter, stir well and bind with some parmesan to taste.

3. Serve the *cotechino* with the purée alongside.

PER PORTION:
816 calories, fiber 0.02 oz/0.5 g

Squash mold

10½ oz/300 g of squash pulp, 2 eggs, 2½ oz/70 g of butter,
2 tbsp of flour, a generous cup of milk,
3½ oz/100 g of gruyere, very finely sliced, salt.

SERVES 6

WINES - **Choose a mature dry white wine with a flowery bouquet, such as Isonzo Malvasia Istriana, Chardonnay di Miralduolo, Saline or Bianco Allavam di Agrigento.**

METHOD

1. Cut the squash into small pieces and boil in salted water for 15 minutes.
2. Separately, prepare a thick béchamel with the flour, butter and milk. Season with salt and add the cheese; stir thoroughly.
3. Separate the eggs, reserving the yolks and whipping the whites until they are stiff.
3. Drain the squash in a sieve and add it to the béchamel. Then add the egg yolks, one at a time, and the whipped egg whites.
4. Grease a baking tin, turn the mixture into it and bake for 30–40 minutes in a moderate oven, about 355°F / 180°C.

PER PORTION:

254 calories, fiber 0.02 oz/0.6 g

Squash loaf

*2 lb 3 oz/1 kg of squash, a generous 1lb/500 g of fresh
mushrooms, the yolks of 3 eggs,
4 tbsp of ricotta, grated parmesan, 1 clove of garlic, oil, salt.*

SERVES 6

WINES -**The rather curious flavor of this dish demands a mature
dry white wine, such as Trentino Pinot Grigio, Colli Piceni, Colli
Martani Grechetto or Castel del Monte.**

METHOD

1. Cook the squash in salted water, then drain, remove the rind, and
pass through a vegetable mill.

2. Clean, slice and fry the mushrooms with the garlic. Add them to the
squash.

3. Return the mixture to the heat to bring out the flavor. Add the ricotta,
parmesan and yolks of egg, and stir until the mixture is smooth.

4. Preheat the oven to 355°F/180°C. Grease a baking tin with oil and
line with sufficient greaseproof paper to cover the tin. Pour the mixture
into the tin, fold over the paper to cover, and bake in the oven for 30–40
minutes.

PER PORTION:

188 calories, fiber 0.09 oz/2.6 g

Squash and spinach roll

1½ cups/200 g of flour, 1 egg, milk, 2 lb 3 oz/1 kg of squash,
1 lb/500 g of spinach, the grated rind of 1 lemon,
3½ tbsp of grated parmesan, 10½ oz/300 g of ricotta,
6 leaves of sage, 3½ tbsp of butter,
nutmeg, salt, pepper.

SERVES 6

WINES - **This dish has a strong sweet taste with an aroma of herbs and spices. Choose a mature, dry white wine, such as Monsupello, Lison-Pramaggiore Pinot, Epomeo or Castelvecchio from the province of Trapani.**

METHOD

1. Work the flour into a dough with the egg and milk and knead for a few minutes, then wrap it in a cloth and leave it to rest in the refrigerator.
2. Bake or steam the squash, leave it to cool, then remove the flesh and drain it in a sieve.
3. Transfer the squash to a saucepan. Add the lemon rind and season with nutmeg, parmesan and salt. Place the saucepan in the refrigerator.
4. Separately, chop the spinach, toss it in a frying pan with the butter. Add the ricotta and parmesan, and season with nutmeg, salt and pepper.
5. Roll out the dough into a rectangle about 12 by 8 in/25 by18 cm or, if preferred, two rectangles half this size. Spread the dough with the squash, followed by the spinach, taking care to avoid the edges, and roll it up lengthwise. Wrap it in a white cloth, secure it with string and boil for 15 minutes in salted water.

6. Remove the roll and chill in the refrigerator. Then cut it into slices, arrange them in an oven dish and dress them with butter, sage and some grated parmesan. Brown them in the oven for a few minutes and serve hot.

PER PORTION:
361 calories, fiber 0.1 oz/2.8 g

Dried cod with onion and squash

2 lb 3 oz/1 kg of squash, rind and seeds removed and diced,
1 onion, chopped, 1 cup/200 g of tomato pulp,
10 oz/300 g of dried cod, soaked and cleaned, flour,
nutmeg, ginger, 1 stock cube, lard,
2 tbsp of olive oil, parmesan.

SERVES 4

WINES - **In this dish the dominant taste of the cod is balanced by the sweetness of the squash and the aroma of the spices. The accompanying wine should be white, well structured, not too young and with a well developed bouquet, such as Colli Berici Riesling Renato, Collio Tokay, Vernaccia di San Gimignano or Castel del Monte.**

METHOD

1. Dip the cod in flour and fry in the lard until golden. Drain and dry on kitchen paper.
2. Separately, lightly fry the onion in the olive oil, add the tomato pulp and the squash, and continue cooking until golden. Add the cod. Crumble the stock cube into the mixture, season with nutmeg and ginger and finish cooking over a low heat. Sprinkle with grated parmesan.

PER PORTION:

123 calories, fiber 0.06 oz/1.8 g

Note: *This recipe, from Villastrada, has been passed down from mother to daughter and has never previously been recorded.*

Fancy squash biscuits

3 cups/350 g of plain white flour, 1¼ cups/150 g of cornflour,
5 oz/150 g of squash, baked, mashed and cooled, the yolks
of 2 eggs, beaten, the white of 1 egg, 1¼ cups/175 g of butter,
¾ cups/200 g of sugar, 1 sachet of baking powder,
1 small glass of sambuca or other sweet liqueur, lard.

SERVES 6

WINES - The consistency of the biscuits and their decidedly sweet flavor steer the choice of an accompanying wine toward a sweet young white with a fruity, fragrant bouquet, such as Loazzolo, Colli Orientali del Friuli Picolit or Moscato di Trani Liquoroso.

METHOD

1. On a pastry board, heap the flour and cornflour. Rub in the butter. Make a well in the center and add all the other ingredients (except the white of egg and the lard) and work into a smooth, fairly firm dough.
2. Roll the dough out to a thickness of about ½ in (1 cm). With fancy biscuit cutters, cut the dough into shapes and arrange them on a baking sheet greased with lard. Brush them with lightly whisked egg white and sprinkle with sugar.
3. Bake at 355°F/180°C for 20 minutes until golden. Allow to cool before serving.

PER PORTION:
221 calories, fiber 0.01 oz/0.4 g

Squash cake

2 lb 3 oz/1 kg of squash, ⅔ cup of milk,
7 tbsp/50 g of almonds, ½ cup/100 g of sugar, 3 eggs,
separated, 3½ tbsp of butter, 10 macaroons, 1 tbsp of bitter
cocoa, 1¼ cups/150 g of cornflour, 1 sachet of baking
powder, vanilla flavoring, salt, butter and breadcrumbs.

SERVES 6

WINES - This classic recipe may be accompanied by young, sweet white wine, such as Torcolano di Breganze, Colli Piacentini Malvasia or Frascati Cannellino.

METHOD
1. Remove the rind and seeds from the squash and cut it into cubes. Place the squash in a covered pan and simmer in the milk, stirring and breaking the squash up with a fork until it has absorbed all the milk.
2. Simmer for 5 minutes, stirring constantly until the squash becomes dry. Allow to cool.
3. Meanwhile, finely chop the almonds and macaroons and add the sugar, cocoa, baking powder, vanilla flavoring, egg yolks, the butter and the squash. Mix well.
4. Whisk the egg whites until stiff and fold into the mixture a little at a time, with a pinch of salt.
5. Preheat the oven to 355°F/180°C. Butter a cake tin and sprinkle it with breadcrumbs. Spoon in the mixture and bake for 1 hour.
6. Allow the cake to cool before serving.

PER PORTION:
449 calories, fiber 0.09 oz/2.6 g

Squash biscuits

13 cups/1.6 kg of plain white flour, 6 cups/750 g of
cornflour, the yolks of 6 eggs,
the whites of 2 eggs, 2 lb 11 oz/1.2 kg of squash,
3¼ cups/700 g of butter, 3¼ cups/800 g of sugar, 1 small glass
of sambuca, 2 sachets of baking powder.
To coat the biscuits: the whites of 4 eggs, caster sugar.

Makes about 100 biscuits

Wines - **These biscuits can be enjoyed with a sweet or medium sweet white wine with an intensely flowery bouquet, such as Erbaluce di Caluso passito, Colli Orientali del Friuli Verduzzo di Ramandolo or Moscato di Pantelleria.**

METHOD

1. Soften the butter. Heap the flour on a board and gradually add all the other ingredients. Work the dough until it has an elastic and even consistency. Leave it to rest for about 20 minutes.

2. Cut it into fingers, arrange them on a buttered baking tray, brush them with the stiffly whisked egg whites and sprinkle them with caster sugar.

3. Bake at 300°F/150°C for about 30 minutes until golden brown.

PER BISCUIT
80 calories, fiber 0.02 oz/0.5 g

Spiced squash cake

1 lb 5 oz/600 g of squash, 2½ cups/300 g of plain white flour,
1 cup/250 g of sugar, 1⅓ cups/200 g of sultanas, salt, 3 eggs,
4 tbsp of olive oil, 2 tsp of cloves, 2 tsp of cardamom,
2 tsp of cinnamon, 1 sachet of baking powder,
1 tbsp of icing sugar to finish.

SERVES 8 to 10

WINES - **A suitable accompaniment to this cake could be either a white or red dessert wine, such as Moscato Rosa dell'Alto Adige, Vin Santo del Chianti or Girò di Cagliari.**

METHOD

1. Butter a cake tin. Cut the squash into slices about ¾ in/2 cm thick and arrange them in the tin. Bake at 355°F/180°C for about 15 minutes until the squash is very soft, then pass it, still hot, through a vegetable mill.
2. Grind the cloves, cardamom and cinnamon to a fine powder. Sieve together the flour, salt and baking powder and mix in the spices.
3. Separately, beat the eggs with the sugar, add the milled squash, the sultanas and the oil and mix well. Then gradually add the spiced flour.
4. Preheat the oven to 355°F/180°C. Butter a fresh cake tin, fill it with the mixture, level the surface and bake for about 1 hour. Leave to cool then sprinkle with icing sugar.

PER PORTION:
349 calories, fiber 0.1 oz/2.9 g

Squash zuccotto

For the cream:

4 cups/1 l of milk, 2 lbs 3 oz/1 kg of squash, boiled and cooled, ½ cup/100 g of sugar, 1¼ cups/150 g of ground almonds, 1¼/150 g of crushed macaroons, 1 tbsp of cornflour, 1 sachet of saffron, ground cinnamon.

To line the mold:

trifle sponge, 1 glass of amaretto liqueur, 1 glass of milk.

For the topping:

3½ oz /100 g of fondant chocolate, 2 or 3 tbsp of cooked and sieved squash.

Serves 10

METHOD

1. Put all the ingredients for the cream into a saucepan and heat gently, stirring until the mixture is well amalgamated and has a creamy consistency. Leave to cool.

2. Separately, soak the trifle sponge in a mixture of milk and amaretto liqueur, and arrange it round the sides of a circular glass mold.

3. Pour the cream into the mold and cover with a layer of soaked trifle sponge. Chill in the icebox for about 2 hours.

4. Melt the chocolate in a double boiler and mix it with the sieved squash. Turn the *zuccotto* out onto a circular serving dish and cover with the chocolate and sieved squash.

PER PORTION:

452 calories, fiber 0.08 oz/2.4 g

WINES - **The accompanying wine for this dessert should be a sweet white with an intense bouquet, such as Trentini Vin Santo, Marsala Oro Sueriore Riserva or Vernaccia di Oristano Liquorosa.**

Squash strudel

*1 lb 5 oz/600 g of squash, rind and seeds removed,
⅓ cup/50 g of sultanas, 1 cup/200 g of sugar, 4 tbsp/30 g of
pine kernels, 4 tsp of butter, 1 cooking apple, 1 tsp of
cinnamon, the grated rind of half a lemon, 2 tbsp of lemon
juice,the yolk of 1 egg, 1 roll of frozen flaky pastry.*

SERVES 4

WINES - **For this classic dessert choose young, sweet white wine
with an intense aroma, such as the sweet yellow Alto Adige
Moscato, the dessert wine Erbaluce di Caluso, and the naturally
sweet Malvasia delle Lipari.**

METHOD

1. Cut the squash and the apple into cubes and simmer for 15 minutes, stirring constantly until the mixture is dry. Allow to cool.
2. Add 2 tbsp of sugar, the lemon rind and lemon juice. With a fork mash away any remaining lumps of squash or apple. As soon as the mixture has cooled, add the sultanas, the pine kernels, the remaining sugar and the cinnamon.
3. Preheat the oven to 355°F/180°C. Open the roll of flaky pastry, spread the mixture over it and dot with the butter.
4. Roll up the pastry, place it in a buttered baking tin and brush it with the egg yolk. Bake for 40–45 minutes. Serve cold.

PER PORTION:
630 calories, fiber 0.12 oz/3.3 g

Squash ice cream

14 oz/400 g of cooked squash, 1¼ cups of milk,
½ cup/125 g of sugar, ½ a vanilla bean, 7 macaroons,
2 tbsp of maraschino.

SERVES 6

WINES - **Because it is served frozen and contains ingredients incompatible with wine, this dessert is best enjoyed without wine.**

METHOD

1. Remove the rind and seeds from the squash, cut into cubes and heat gently with the milk and the vanilla bean.
2. Add the sugar, allowing it to dissolve thoroughly. Leave to cool for a few minutes.
3. Whisk the mixture, taking up the thick liquid at the bottom of the pan. Add the macaroons and the liqueur. When it has cooled, transfer the mixture to an ice cream maker.

PER PORTION:

205 calories, fiber 0.035 oz/1 g

Squash recipes by famous Italian chefs

"CECILIO, KING OF THE SQUASHES, SPLITS THEM, CUTS THEM INTO A THOUSAND SLICES, AS THOUGH THEY WERE THE SONS OF TIESTE.
YOU SHALL EAT THEM FIRST OF ALL IN THE ANTIPASTO; HE WILL PRESENT THEM TO YOU IN THE FIRST AND SECOND COURSE AND IN THE THIRD;
WITH THEM HE WILL PREPARE THE FINAL COURSE FOR YOU. WITH THEM THE PASTRY COOK WILL MAKE YOU SOME SWEET-TASTING BUNS, SOME LITTLE
CAKES OF VARIOUS KINDS AND SOME DATES WELL KNOWN IN THE THEATERS. WITH THEM THE CHEF WILL PREPARE VARIOUS DELICIOUS DISHES,
SUCH THAT YOU COULD BELIEVE THAT HE HAD USED LENTILS AND BEANS; HE IMITATES MUSHROOMS AND SAUSAGES, TAILS OF TUNA AND LITTLE SARDINES.
WITH THESE THE HEAD CHEF DISPLAYS HIS ABILITY."*

The recipes that follow demonstrate the culinary versatility of the squash, and show how today's chefs, not unlike Cecilio's cook and head chef, use them to prepare a great variety of delicious dishes, from the antipasto to the dessert, with results that would have satisfied even Martial.

*Martial, (c. 40–c.104) *Epigrams*

Crawfish with yellow squash

AND SPINACH

4 lb 7 oz (2 kg) of crawfish, 1 bunch of fresh herbs,
including thyme, parsley, celery and tarragon,
1 glass of dry white wine, the yolks of 6 eggs,
¾ cup of cream, 3 handfuls of spinach leaves, butter, olive oil,
1 tsp of tomato concentrate, 7 oz/200 g of yellow squash,
rind and seeds removed, salt.

SERVES 4

WINES - **The particularly delicate taste of the crawfish, the pleasantly sweet nature of the squash and the fat of the cream, together with the fragrant aromas of the herbs included in the mixture for this recipe, make the ideal accompaniment to this dish a young dry white wine, with a fairly intense bouquet, such as Alto Adige Veltliner, Colli Bolognesi Sauvignon or Menfi Bianco Feudo dei Fiori.**

METHOD

1. Cook the squash in the oven and then sieve it.

2. Add the wine and herbs to 8–10 pt (4–5 l) of salted water and boil for at least 5 minutes. Throw in the crawfish and cook for 3 minutes. Drain, cool under cold water and shell them.

3. Toss the raw spinach leaves in oil, butter and salt. Drain them and keep them hot.

4. Whisk together the cream, squash purée, egg yolks, tomato concentrate, salt and pepper. Gently heat the mixture in a thick-bottomed pan, stirring constantly with a wooden spatula until it thickens.

5. Pour the mixture into an oven dish and arrange the crawfish and spinach leaves on top. While the crawfish will sink down into the mixture, the spinach will remain on the surface, creating an attractive color effect. Brown under the grill for 2–3 minutes.

PER PORTION:

533 calories, fiber 0.035 oz/1.1 g

Whisked squash Ippolito Cavalcanti

1 lb 12 oz/800 g of squash,
¼ cup/80 g of bread soaked in milk,
3 tbsp/20 g of breadcrumbs, 4 eggs, separated,
3½ oz/100 g of grated mild mozzarella.

For the sauce:
3 oz/70 g of fontina, the yolk of 1 egg, 2 tsp of milk,
1 bunch of parsley.

SERVES 4

WINES - **To accompany this dish choose light, young, red wine, such as Friuli-Grave Merlot or the Cilento.**

METHOD

1. Preheat the oven to 355°F/180°C and bake the squash until it is soft.

2. Pass it through a fine sieve.

3. Add the bread, the grated cheese and the 4 beaten egg yolks, and season with salt.

4. Whisk the egg whites until stiff and fold them into the mixture.

5. Butter a mold large enough for the mixture to fill it only halfway up and place in a larger oven-proof vessel filled with water. Sprinkle the mold with breadcrumbs and pour in the mixture. Bake at 355°F (180°C) for 20–25 minutes.

6. To make the sauce, melt the cheese in the milk over a gentle heat and, when tepid, add the egg yolk and the parsley, finely chopped.

7. Pour the sauce into a serving dish and carefully empty the squash mold onto it.

PER PORTION:
324 calories, fiber 0.06 oz/1.6 g

Note: *This recipe is taken from "Cucina teorica pratica" by Ippolito Cavalcanti, Duke of Buonvicino, who, in the 19th century, while staying at Napoleon's court, collected and arranged recipes from southern Italy. Cavalcanti also suggests this recipe in a sweet version, with the addition of sugar and, of course, without the cheese.*

Fried squash alla siciliana

1 lb 12 oz/800 g of squash, rind and seeds removed,
2 glasses of oil, 1 clove of garlic,
½ a glass of vinegar, salt,
about 20 black Sicilian olives,
fresh mint to garnish.

SERVES 4

METHOD

1. Cut the squash into cubes and soak in salted water for about 2 hours.
2. Drain and dry the squash, then fry it in very hot oil. Lift it out of the frying pan and set it aside.
3. Add the garlic to the pan and lightly fry it. Then return the squash to the pan and leave it for about 5 minutes to absorb the flavor.
4. Lightly fry the olives and add them to the squash. Garnish with the mint just before serving.

WINES - As a accompaniment for this dish, with its strong flavors, choose young dry white wine, possibly slightly sparkling, such as Colli Berici Garganega or a sparkling Soave.

PER PORTION:
101 calories, fiber 0.07 oz/2 g

Soused squash

1 lb 12 oz/800 g of squash,
6 cloves of garlic,
4 tsp of parsley, salt,
flour, peanut oil for frying

SERVES 4

METHOD

1. Slice the squash into strips. Place in a colander, sprinkle with salt and leave for 2–3 hours under a weight so that it will lose water.

2. Heat plenty of peanut oil in a frying pan.

3. Lightly flour the squash and fry it in the oil. When it is golden on both sides drain it well on kitchen paper.

4. Finely chop the garlic and parsley and mix them with the vinegar.

5. Arrange the squash in a generously large oval dish and pour over the vinegar, garlic and parsley sauce.

6. Leave to marinate for at least 4–6 hours, then serve cold or warm, either as a starter or as a side dish.

Wines - This unusual dish, which is generally served as a starter, has a balanced flavor and aroma. The accompanying wine should be a young white with a fruity bouquet, such as Oltrepò Pavese Cortese or Bianco Vergine Valdichiana.

PER PORTION:
82 calories, fiber 0.04 oz/1.3 g

Squash tart

ALLA QUISTELLESE

4 thin slices of squash, each weighing about 1½ oz/40 g,

2½ cups/300 g of parmesan, flaked,

3½ tbsp of butter,

½ a glass of cooking oil,

fortified wine.

SERVES 4

METHOD

1. Fry the squash in plenty of hot oil then drain them on kitchen paper.

2. Arrange the squash in a baking dish, with the parmesan and a few knobs of butter.

3. Place it on an oven tray and bake at 250°F/120°C for 7 minutes until golden.

4. Serve very hot, with the addition of a little fortified wine to taste.

WINES - **This simple but appetizing dish, with its well-balanced flavors, requires a young dry white wine with a fairly intense fruity bouquet, such as Lugana or Lison-Pramaggiore Verduzzo.**

PER PORTION:

397 calories, fiber 0.01 oz/0.3 g

Squash tarts

WITH BLACK TRUFFLES

5¼ oz/150 g of filo pastry, 10½ oz/ 300 g of squash,
rind and seeds removed, 3 tbsp of grated parmesan,
1½ tbsp of roughly chopped walnuts,
2 tsp of chopped shallot, a generous tsp of butter,
1 black truffle, ½ a ladle of stock, salt and pepper,
extra virgin olive oil, parsley to garnish.

SERVES 6

WINES - **A suitable accompaniment for this dish, with its delicate combination of flavors, would be a young dry white wine with a fairly intense, flowery bouquet, such as Terlano Pinot Grigio or Biancolella d'Ischia.**

METHOD

1. Roll the pastry out thinly and cut out 24 disks of even size. Arrange them on an oven tray, weighed down with dried beans, bake them at 250–265°F/120–130°c for about ¼ hour, until they are crisp and golden.

2. Cook the squash in a fan assisted oven, or microwave, until it is soft. If necessary cover it with greaseproof paper to prevent the surface from drying.

3. Pass the squash through a vegetable mill and add the grated parmesan.

4. Melt the butter in a small frying pan, add the shallot and cook until soft. Add the chopped truffle. Slake with the stock, reduce it a little, and add the squash mixture. Season with salt and pepper to taste and add the walnuts.

5. To make each tart, take a pastry disk, add a spoonful of squash, and repeat until you have four disks of pastry and three layers of filling. Place each tart on a plate, dress it with a little olive oil, garnish with a sprig of parsley and serve warm.

PER PORTION:
198 calories, fiber 0.03 oz/0.9 g

Marinated squash

1 lb 6 oz/600 g of squash, the rind and seeds removed, 3 tbsp of extra virgin olive oil, 2 tsp/10 g of cooking salt, black pepper, 4 dried bay (or basil), 2 peppermint leaves.

For the marinade:

⅔ cup of red wine, 7 tbsp/100 ml of wine vinegar, 2 tbsp of sugar.

SERVES 6

WINES - **The strong flavor of the vinegar and the peppermint in this dish suggests that it is best enjoyed without wine. However, if a wine is to be served, an aperitif wine, the same as that drunk before the meal, can be offered.**

METHOD

1. Cut the squash into evenly sized pieces and fry them in a non-stick pan in the olive oil, turning them so that they are perfectly cooked. Season with salt and pepper and continue to cook on a moderate heat for a few more minutes until they have formed a golden skin.

2. Drain the squash on kitchen paper, then put it in an oven dish and arrange the bay and mint (or basil) leaves on top.

3. Mix the wine, sugar and vinegar in a saucepan, bring the liquid to a boil and pour it over the squash. Tightly cover the pan and leave to rest for at least 3 hours.

4. Serve dressed with olive oil.

PER PORTION:

126 calories, fiber 0.03 oz/0.8 g

Squash alla murgese

2 lb 3 oz/1 kg of squash, 2 lb 3 oz/1 kg of mushrooms
(ideally cardoncelli from Murgia, Bari),
7 tbsp of extra virgin olive oil, 2 cloves of garlic, crushed,
salt and pepper, oregano and parsley.

SERVES 6

METHOD

1. Peel and dice the squash, clean the mushrooms and slice them fairly thickly.

2. Fry the garlic in the oil, and as soon as the garlic is browned add the squash and the mushrooms. Season with salt and pepper and cook together for about 20 minutes.

3. Chop the parsley, oregano and a clove of garlic together and add immediately before serving.

WINES - **This simple dish, with a pleasingly rustic character, requires the accompaniment of a moderately mature white wine, such as Castel Monte or Montecarlo.**

Note: *This recipe can be served as a side dish. With the addition of 1¼ cups/300 g of liquidized tomatoes it can be used to dress pasta.*

PER PORTION:
183 calories, fiber 0.15 oz/4.3 g

Cresc Tajat

WITH RABBIT AND HAIRY SQUASH

For the sauce:
2 rabbit thighs, 4 tomatoes, 10 leaves of fresh basil,
1 shallot, 7 tbsp of virgin olive oil,
14 oz/400 g of hairy squash.
(This is a globular squash the size of an orange, with a green
hairy rind and whitish flesh.)

For the cresc tajat (polenta gnocchi):
3 oz/60 g of polenta, 1⅓ cups/160 g of wheat flour, water,
1 tbsp of parmesan.

SERVES 4

WINES - This dish should be accompanied by a young red wine with a fairly intense bouquet, such as Teroldege Rotaliano or Oltrepò Pavese Pinot Nero.

METHOD

1. Remove the bones and veins from the rabbit thighs, peel the tomatoes, and remove the rind and seeds from the squash. Dice all these ingredients.

2. Chop the shallot and brown it in a non-stick frying pan. Add the diced rabbit, tomato, and squash, season with salt and pepper, then add the finely chopped basil at the last minute.

3. Work the polenta and flour together, form the dough into a number of long, thin rolls, put them one on top of another and slice them to make small, irregular gnocchi-like shapes.

4. Cook the *cresc tajat* in boiling salted water, toss them in the frying pan with the sauce, sprinkle with parmesan, and serve hot, garnished with basil leaves.

PER PORTION:
453 calories, fiber 0.14 oz/4.1 g

Squash and potato pasties

ALLA LASTRA

3½ oz /100 g of potatoes, 3½ oz/100 g of squash,
¾ tsp of garlic, 1½ oz/40 g of bacon, 2 eggs, beaten,
⅓ cup/40 g of parmesan.
For the pasta:
3 cups/ 400 g of flour, salt, water.

SERVES 4

METHOD

1. Boil the potatoes in their skins, then peel and mash them.

2. Remove the rind and seeds from the squash and cut it into pieces. Bake at 355°F/180°C until soft, then mash it.

3. While the squash is baking, finely chop the garlic and bacon and fry them lightly.

4. Combine the potato, squash, salt, parmesan, and eggs with the fried garlic and bacon.

5. Make a sheet of pastry with the flour and water and cut it into 4 in/10 cm squares.

6. Fill the squares with the prepared mixture, fold them and seal the edges well.

7. Put a stone or terracotta slab on the fire and when it is hot cook the pasties on it as though they were pitta bread. Alternatively they can be cooked on an oven tray. Serve hot.

WINES -**This dish, with a discreet flavor of garlic and bacon, should be accompanied by a mature white wine, such as Roero Arnels or Colli Bolognesi Chardonnay.**

PER PORTION:
526 calories – fiber 0.17 oz/4.9 g

Squash soup

WITH TOASTED ALMONDS

¼ cup/600 g of squash, rind and seeds removed,
1¾ oz/50 g of butter, 2 onions, 1 glass of fresh cream,
4 tbsp of chopped toasted almonds, the yolk of 1 egg,
6 cups of capon or chicken stock,
salt and pepper.

SERVES 4

WINES - **This is a traditional Mantuan soup, delicate and pleasing. The accompanying wine should be sweet and light, such as Tokay of San Martino della Battaglia or Trentino Pinot Bianco.**

METHOD

1. Finely chop the onions and add them to melted butter in a frying pan. Cook over a low heat until soft.

2. Cut the squash into strips and add it to the onions. Season with salt and pepper. Continue to cook on a low heat for a few minutes, then add boiling stock until the desired consistency is reached. Cook on a low heat for a further 25–30 minutes then liquidize in a blender.

3. Return to the heat and bring to a boil. If the soup becomes too thick, add more stock.

4. Remove the soup from the heat. Mix together the egg yolk and the cream, stir into the soup and serve sprinkled with the toasted almonds.

PER PORTION:
53 calorie, fiber 0.11 oz/3.2 g

Squash lasagne

WITH SLICES OF FOIE GRAS

14 oz/400 g of fresh egg lasagne,
a generous pound/500 g of squash, baked and sieved,
a generous pound/500 g of foie gras, the yolk of 3 eggs,
7 tbsp of white wine, 2 tbsp of tarragon vinegar,
½ cup/70 g of grated parmesan, 3½ tbsp of melted butter,
1 oz/30 g of crushed macaroons,
thyme, chervil, fresh marjoram.

SERVES 10

WINES - **A suitable accompaniment to this innovative dish is a dry white wine with a fairly intense, flowery bouquet, such as Terlano Pinot Bianco or Orvieto Classico.**

METHOD

1. Cut the pasta into disks about 5 in/12 cm across, cook in boiling salted water, refresh, drain and spread them out on a clean tea towel.

2. Pour the wine, the tarragon vinegar, the egg yolks and half the parmesan into a double boiler and whisk up over a gentle heat.

3. Heat the sieved squash to remove any remaining water, whisk in a blender and add to the liquid.

4. Cut the foie gras into slices less than ¼ in/3–4 mm thick, dip in flour and season with salt and pepper. In a non-stick pan, fry them lightly on both sides in very hot oil, remove and drain on kitchen paper.

5. Butter 10 molds the same size as the pasta disks. In each, lay a pasta disk, followed by a layer of the creamed squash and a slice of foie gras sprinkled with a little grated parmesan. Repeat so as to make another three layers, finishing with a layer of pasta.

6. Turn out the molds onto a buttered oven dish and sprinkle with parmesan and crumbled macaroon. Heat in the oven for 10 minutes then brown under the grill for 3 minutes.

7. Serve the lasagne on individual plates, garnished with fresh herbs and, if desired, a light grating of white truffles.

PER PORTION:
333 calories, fiber 0.07 oz/1.9 g

Potato and squash gnocchi

WITH FISH STEW

For the gnocchi:

*2 lb 2 oz/1 kg of potatoes, 1¼ cups/150 g of flour,
1 egg, a pinch of nutmeg, salt.*

For the filling:

10½ oz/300 g of yellow squash, rind and seeds removed, salt.

For the stew:

*1 sole, 3½ oz/100 g of shrimps, 5¼ oz/150 g of bream,
5 oz/150 g of scampi, 1 ripe tomato, 1 shallot, 1 bunch of
parsley, ½ a glass of dry white wine, ½ cups/100 g of butter,
2 oz/50 g of black truffle.*

SERVES 6

WINES -**The flavor of the fish stew contrasts perfectly with the
subtle sweetness of the squash and the potato gnocchi. To
accompany this dish choose dry white wine, with a fruity bouquet,
such as Colli Orientali del Friuli Ribolla Gialla or Frascati Superiore.**

METHOD

1. Boil the potatoes and mash them finely. Add the egg, flour, nutmeg and salt, mix well and roll out to a thickness of about ¾ in/1.5 cm. Stamp out rounds with a glass.
2. Bake the squash in the oven, whisk it up and add salt to taste. Reserve a quarter of the squash, and spoon the rest over half the number of potato rounds.
3. Top off with the remaining potato rounds and seal the edges.
4. Clean and scale the fish, cut it into pieces and toss them in a frying pan with half the butter and the chopped shallot. Add salt to taste, bathe with the wine, reduce a little, add the chopped parsley and the diced tomato. Keep warm.
5. Toss the potato rounds in the frying pan with the remaining butter.
6. Pour a little of the remaining squash sauce onto warm plates, then arrange the gnocchi in a circle and spoon the fish stew into the center. Garnish with parsley and strips of truffle.

PER PORTION:
448 calories, fiber 0.25 oz/7.3 g

Squash cream soup

WITH MACAROONS

1 lb 8 oz/700 g of squash, rind and seeds removed,
2 oz/60 g of shallots, ¼ cup/60 g of butter, ¾ cup of chicken
stock, a generous 3 cups of single cream, 1 small tsp of salt,
⅔ tbsp of sugar, 16 macaroons.

SERVES 8

METHOD

1. Chop the shallots and dice the squash.

2. Melt ⅔ of the butter in a saucepan and cook the shallots until soft.

3. Add the diced squash. When it has dried out add the hot stock and cook until the squash is soft.

4. Pass the squash through a vegetable mill and transfer to a fresh saucepan. Add the sugar, salt and cream, then return the mixture to the heat and bring it to a boil. Add the remaining butter and stir.

5. Ladle into warmed soup plates and serve with 2 macaroons to each portion.

WINES - As an accompaniment to this delicate dish choose a young, dry white wine with a fruity bouquet and a hint of nuts and spices, such as the Trentino Traminer Aromatico or the Fiovano Semillon.

PER PORTION:
439 calorie, fiber 0.05 oz/1.4 g

Squash cream soup

WITH ALMONDS AND ORANGES

1 lb 12 oz/800 g of squash, rind and seeds removed,
a generous cup/100 g of ground almonds,
the yolks of 2 eggs, ¼ cup of milk,
¼ cup of beef and chicken stock, salt,
the juice of 5 freshly squeezed oranges.

Serves 4

METHOD

1. Bake the squash and leave it to cool, then pass it through a vegetable mill and transfer it to a saucepan.

2. Add the ground almonds, the egg yolks and the orange juice and whisk together.

3. Over a gentle heat, continue to whisk, gradually adding the milk and the stock so as to obtain a velvety cream.

4. Serve in warmed soup bowls.

WINES - Because of the squash and the almonds, this dish has an intense sweetness and is considerably rich. The accompanying wine should be a young, dry white with a fruity bouquet such as Regaleali Bianco or Breganze Vespaiolo.

PER PORTION:
227 calories, fiber 0.16 oz/4.6 g

Radicchio and ricotta twists

WITH SQUASH SAUCE

For the pastry:

1¼ cups/200 g of flour, 3½ tbsp/50 g of butter,
3½ tbsp/50 ml of water,
the yolks of 2 eggs,
salt and pepper

For the filling:

1 small onion, a generous 5 oz/150 g of radicchio,
3½ oz/100 g of fresh ricotta cheese, the yolk of 1 egg,
½ cup/50 g of parmesan, ½ glass of red wine,
salt and pepper.

For the sauce:

7 oz/200 g of squash, 6½ tbsp/100 ml of olive oil,
white pepper, salt.

SERVES 4

WINES - A suitable wine for this dish would be a dry white, not too young, such as Regaleali Nozze d'Oro or Collio Tokay.

METHOD

1. Combine all the ingredients for the pastry so as to obtain a firm dough. Leave to rest for 30 minutes.

2. Finely chop the onion and fry in olive oil until soft but not browned.

3. Wash the radicchio, slice it thinly and add to the onion. When partially cooked, bathe with red wine and simmer until the radicchio is soft. Add the ricotta, parmesan, egg yolks, salt and pepper.

4. Roll out the pastry and cut it into small rectangles about 3 by 4 inches/ 8 by 12 cm. Place some of the filling on each of them and seal by twisting the corners together.

5. Preheat the oven to 355°F/180°C and cook the twists for 8–10 minutes.

6. To prepare the sauce, boil the squash for a few minutes, drain, and liquidize in a blender. Add the oil, season with salt and pepper and whisk, adding a little water if the mixture is too thick. Return the sauce to the heat.

7. Pour the sauce into a dish, add the twists and garnish to taste. Serve hot.

PER PORTION:
163 calories, fiber 0.03 oz/0.8 g

Pasta with squash

A generous 10oz/300 g of squash, rind and seeds removed,
2 cloves of garlic, 1 medium-sized chili pepper,
7 tbsp of extra virgin olive oil, parsley, basil, salt,
14 oz/400 g of mixed pasta or broken-up spaghetti.

SERVES 4

WINES -This simple dish is best accompanied by a young dry white wine such as Colli Piacentini Ortrugo or white Pomino.

METHOD

1. Crush the garlic and toss it with the oil in a saucepan. When it is nicely browned add the cubed squash, turn down the heat and fry lightly for a few moments. Then add the pepper; cover the pan and allow to cook for 6–7 minutes, stirring from time to time.

2. Add salt to taste, and if the squash has not made enough water add a glassful. After 7–8 minutes check the salt and remove from the heat, keeping the pan warm.

3. Meanwhile cook the pasta in boiling salted water, drain, add to the sauce and serve hot.

PER PORTION:
564 calories, fiber 0.12 oz/3.5 g

Note: *Watery, stringy varieties of squash, especially those from Ischia, are particularly suitable for this dish.*

Cream of squash

WITH STUFFED CHOUX

For the cream:

9 oz/250 g of squash, rind and seeds removed,
5 oz/150 g of leeks, 1 cup of fresh cream,
9 oz/250 g of potatoes, the yolk of 1 egg, salt,
vegetable stock, extra virgin olive oil.

For the choux:

1 cup of water, ¼ cup/60 g of butter, 1 cup/125 g of flour,
3 eggs, 1 cup béchamel , a handful of spinach,
3½ tbsp of parmesan.

SERVES 4

WINES - A suitable accompaniment for this dish would be a dry white wine with an aromatic bouquet, such as Aquileia-Traminer Aromatico or Solopaca Falanghina.

METHOD

For the cream:

1. Finely chop the squash, potatoes and leeks, place them in a saucepan, cover them with the stock, add salt and cook for ½ hour.

2. Liquidize the vegetables in a blender, thinning the mixture with more stock if necessary. Add a little oil and then the cream.

For the choux:

3. Bring the water to a boil with a pinch of salt and the butter. Add the flour and stir vigorously with a wooden spoon to dissolve any lumps. Allow to cool.

4. Add the eggs one at a time, stirring each into the mixture.

5. Preheat the oven to 355°F/180°c. Line an oven tray with greaseproof paper and, using a forcing bag, pipe an evenly spaced amount of the choux mixture onto it. Bake for about 30 minutes.

6. Cook, strain and purée the spinach. Prepare a fairly thick béchamel, adding some parmesan. Divide the mixture into two and add the cooked spinach to one half. Using a biscuit forcer, fill one half of the number of choux with the plain béchamel, and the other half with the spinach béchamel.

7. Arrange the choux on plates, two of each flavor per portion and serve with the cream of squash.

PER PORTION:

651 calories, fiber 0.12 oz/3.5 g

Risotto with squash

AND CHEESE FONDUE AND TRUFFLE

1 lb/500 g of yellow squash, rind and seeds removed,
1 white onion, chopped, ½ cup/100 g of butter,
10½ oz/300 g of cheese fondue,
1½ cups/300 g of Arborio rice, 1 white truffle,
vegetable stock, white wine.

Serves 4

METHOD

1. Chop the onion and fry it lightly in half the butter. Dice the squash, add it to the onion and cook over a low heat for a few minutes. Add the rice, let it brown slightly, bathe it with white wine and allow it to absorb the liquid.

2. Heat the stock and add it to the rice, a little at a time.

3. When the rice is cooked, take it off the heat and mix in the remaining butter.

4. Transfer the rice to a serving dish and sprinkle with the cheese fondue and truffle.

Wines - **For this flavorsome autumn dish choose a young red wine with an intensely fruity, floral bouquet such as Barbera d'Alba or Colli Morenici Mantovani del Garda.**

PER PORTION:
772 calories, fiber 0.12 oz/3.5 g

Squash soup au gratin

1 round squash weighing about 6½ lb/3 kg, 4 leeks,
1⅔ cups of vegetable stock, ¾ cup of cream,
10½ oz/300 g of cottage cheese,
8 slices of bread (preferably homemade),
salt, pepper and oil.

SERVES 6

WINES - The accompaniment for this dish should be a dry white wine, such as Breganze Pinot Grigio or Greco di Tufo.

METHOD

1. Cut out the crown of the squash and remove the seeds and pith. Using a knife, remove the flesh, taking care to keep the rind intact. Set aside the flesh: this should make just under 6 cups/1.3 kg.

2. Trim and wash the leeks, slice them finely and sweat them for a few minutes in a pan with a little oil.

3. Heat the oven to 355°F/180°C. Cut the squash flesh into large cubes, season with salt and pepper and bake it with a little oil for about 20 minutes.

4. When the squash is soft, whisk it up with the leeks.

5. Transfer this mixture to a pan with the vegetable stock and cook for 15 minutes over a low heat. Remove from the heat and add the cream.

6. Finely mash the cottage cheese and toast the bread.

7. Place the hollowed rind of the squash on an oven tray. Cover the bottom with cottage cheese, following with 4 slices of toast and some of the squash mixture. Repeat and finish with a layer of cheese.

8. Bake at 355°F/180°C for 20 minutes and finish off under the grill for a few moments.

9. Bring the whole squash to the table and serve very hot.

PER PORTION:
585 calories, fiber 0.24 oz/6.9 g

Squash tortelli

WITH A LIGHT HARE STEW

14 oz/400 g of yellow squash, rind and seeds removed,
3 tbsp/20 g of parmesan, ¼ cup of extra virgin olive oil,
salt and pepper.

For the pasta:
2 eggs, 2 cups/250 g of flour, 2 tsp/10 ml of oil, salt.

For the stew:
1 hare's thigh, 2 oz/50 g of onion, 1½ oz/40 g of leek,
¼ oz/20 g of carrot,
¼ oz/20 g of celery, half a bay leaf, 2 cloves,
1 stick of cinnamon, ¼ oz /10 g of black pepper,
1 ripe tomato, 2 cups/½ liter of red wine,
7 tbsp of extra virgin olive oil,
salt and pepper.

SERVES 4

WINES - **An appropriate wine for this dish would be a mature red, dry, fruity and full-bodied, such as Franciacorta or Colli Piacentini Cabernet Sauvignon.**

METHOD

1. Marinate the hare overnight in the wine and with all the other ingredients for the stew except the tomato.

2. Bake the squash with salt, pepper and olive oil. Press it through a sieve, collecting the flesh in a basin. Add the parmesan and oil, and add salt and pepper to taste.

3. Mix and roll out the pasta dough and cut it into oblongs. Fill these with squash, close them up to form *tortelli* (pasties) and place in the refrigerator.

4. Drain the hare and the vegetables. Toss the vegetables in olive oil. Fry the hare separately. Combine them in a pan and add the wine and tomato. Bring to a boil, removing any scum that forms on the surface and simmer for about 40 minutes.

5. Remove the hare, take the flesh off the bone and dice it. Pass the vegetables through a mill, and adjust the seasoning. Return the hare meat to the vegetables and keep warm.

6. Boil the *tortelli* in salted water until cooked. Dress with the hare stew and serve.

PER PORTION:
1080 calories, fiber 0.11 oz/3.1 g

Squash soup

WITH RICE AND CHEESE AND CINNAMON CRISPS

1 lb/500 g of squash, rind and seeds removed, 1 rib of celery,
1 large potato, 2 carrots, 1 onion, 2 cloves of garlic,
⅓ cup/50 g of butter, 4½ tbsp of extra virgin olive oil,
1 cup/100 g of pilaff rice, cooked, salt, pepper, cinnamon,
1 or 2 ladles of chicken stock,
3 tbsp/20 g of grated parmesan.

For the crisps:
6 tbsp/40 g of grated parmesan, powdered cinnamon.

SERVES 4

WINES -**The accompanying for this fragrant soup wine should be a young dry white, such as the Bianco di Custoza or the Est! Est!! Est!!! di Montefiascone.**

METHOD

1. Slice the vegetables and fry them in the oil and half the butter on a low heat for about 10 minutes.

2. Dice the squash and add it to the vegetables. Let it brown a little, season with salt and pepper, cover with chicken stock, and cook for 1 hour. Liquidize in a blender.

3. Add the pilaff rice and bring the liquid to a boil. Bind and flavor it with the parmesan, the remaining butter, and a small pinch of cinnamon.

4. To prepare the crisps, mix the parmesan with a little cinnamon. Drop patches of the mixture into a hot frying pan. As they melt take them from the pan and give them a curved shape by resting them over a rolling pin.

5. Serve the soup in warm plates, decorating each with a crisp.

PER PORTION:
564 calories, fiber 0.26 oz/7.5 g

Pasta roll

WITH POTATOES, SQUASH AND MUSHROOMS

For the pasta:
1½ cups/175 g of flour, 2 eggs, truffles

For the filling:
7 oz/200 g of potatoes, 10 spinach leaves,
7 oz/200 g of squash, rind and seeds removed,
5 oz/150 g of mushrooms, 1 clove of garlic, 2 eggs,
rosemary and parsley,
1 cup/100 g of parmesan.

For the squash sauce:
7 oz/200 g of squash, 1 oz/30 g of onion, 3 macaroons,
4 cups/1 l of stock, balsamic vinegar, truffle.

SERVES 4

WINES - **This modern recipe demands a young dry white wine, such as Colli Bolognesi Riesling or Garda Orientale Sauvignon.**

PER PORTION:
577 calories, fiber 0.25 oz/7.3 g

METHOD

1. Heap the flour, make a well in the center, and add the eggs, then work rapidly into a smooth dough. Leave it to rest for 4 hours. Roll it out to a thickness of ⅛ inch/3 mm and divide into rectangles 12 x 8 in/30 x 20 cm.

2. Steam the potatoes for 20 minutes. Peel and sieve them, then mix with the chopped herbs and the eggs.

3. Bake the squash at 355°F/180°C for 20 minutes. Pass it through a fine sieve, season with salt and leave to cool.

4. Wipe the mushrooms and remove the stalks. Slice them and toss them in the frying pan with a clove of garlic. Season with salt and sprinkle with some chopped parsley. Parboil the spinach.

5. Lay out the rectangles of dough, spread the potato purée over them, followed by a layer of spinach leaves and mushrooms, finishing with the squash purée. Sprinkle with parmesan and roll the rectangles up. Wrap each roll in a cloth, tie the ends and steam them in a pressure cooker for 20 minutes.

6. Allow the rolls to cool, then unwrap them, roll in parmesan, cut into slices and brown in a non-stick pan or under the grill.

7. To make the squash sauce, chop the onion and sweat it in the butter, add the macaroons and the diced squash, cover with the stock, cook for 1 hour on a low heat, then pass the sauce through a fine strainer.

8. Pour the sauce into the center of each plate, place a slice of the roll on the sauce and garnish with truffle and two drops of balsamic vinegar.

Risotto with squash

AND RED WINE

1½ cups/320 g of superfine Carnaroli rice,
2 tsp/10 ml of white wine, 14 oz/400 g of yellow squash,
rind and seeds removed, 2 tbsp/15 g of pine kernels,
1 tsp/5 g of sugar, 1 tsp/5 ml of red wine vinegar,
3½ tbsp/50 ml of extra virgin olive oil,
4 cups/1 l of vegetable and/or meat stock,
½ a white onion, trimmed, 1⅔ cups of red wine,
3½ oz/100 g of butter,
3½ tbsp/50 g of grated parmesan.

SERVES 4

WINES - **A suitable accompaniment for this innovative dish with its rich and complex flavors would be a red wine, medium sweet, sweet or even sparkling, with a fairly intense bouquet, such as Oltrepò Pavese Sangue di Giuda, or Freisa di Chieri Amabile.**

METHOD:

1. Pour the red wine into a pan and reduce it to ¾ of its initial volume.
2. Separately, in a stainless steel pan, gently heat the olive oil, add the sugar and let it caramelize for a few minutes. Add the diced squash and the pine nuts and lightly fry for 4–5 minutes. Then add the vinegar followed by half the white wine, making both steam. Continue cooking gently until the squash is half cooked.
3. In a fresh stainless steel pan, melt the butter and brown the rice, then quench it with the remaining white wine. Let the wine evaporate well then cover the rice with the stock and bring it to a boil. When it has been boiling for 5 minutes add the squash.
4. Finish cooking the rice (about 18 minutes in all), then add the butter and parmesan and cook until creamy, taking care to keep the rice firm.
5. Pour into the dishes, sprinkle with the reduced red wine and serve.

PER PORTION:
760 calories, fiber 0.07 oz/2 g

Milled squash and fresh ricotta

1 lb 1 oz/500 g of squash, with the rind,
3½ oz/100 g of fresh goat's milk ricotta,
10 squash flowers (petals only),
1 grated carrot, 1 tbsp of chopped shallot,
½ a cup of sweet Albana di Romagna,
about 1 cup of white meat broth,
1 tbsp of grated parmesan, 8 tsp/40 g of butter,
nutmeg, salt.

SERVES 4

WINES - An excellent accompaniment to this dish is a young sweet white wine, such as a Torgeltropfen from Alto Adige, an Albana from Romagna, a Tacelenghe from Buttrio, a medium sweet Orvieto or a Torricella Malvasia.

METHOD

1. Preheat the oven to 355°F/180°C and bake the squash for about 20–25 minutes, or until it has dried out a little. Remove the rind and any burnt bits and slice the rest.

2. Soften the shallot and carrot in the butter over a moderate heat, then add the sliced squash flowers and the squash. Cook for 10 minutes. Add the wine and allow it to reduce, then remove from the heat.

3. Pass the vegetables through a sieve or vegetable mill (using the finest shredder), add the ricotta, blending it well, then a little nutmeg, the parmesan, and the broth, taking care not to make the mixture too dilute. Season with salt to taste.

4. Heat the mixture in a double boiler and when it is very hot pour in warmed bowls.

PER PORTION:
169 calories, fiber 0.05 oz/1.5 g

Wholemeal gnocchi

WITH BEANS AND SQUASH

1 lbs 3 oz/1 kg of floury potatoes,
3¼ cups/400 g of wholemeal flour, 1 lb/500 g of fresh beans,
1 lb/500 g of squash, 2 tbsp extra virgin olive oil, salt,
sage, 1 clove of garlic, rosemary, butter,
2 cups/½ l of vegetable stock,
2 cups/½ l of milk.

SERVES 6

METHOD

1. Boil the potatoes. Mash them and mix with the flour. Shape the mixture into small gnocchi, about the size of a butter bean.

2. Brown the beans with a knob of butter and the sage. After a few minutes cover them with stock and leave them to cook for ½ hour.

3. Peel and cut up the squash. Brown it in the olive oil, with the garlic and rosemary, pass it through a vegetable mill. Add salt to taste.

4. Boil the gnocchi in plenty of salted water, drain them and put them in a tureen with the beans and mix them with a spoonful of extra virgin olive oil.

5. Spoon the creamed squash onto plates, followed by a generous spoonful of gnocchi and beans, topped off with a twist of black pepper.

WINES -**The accompanying wine should be a young red, such as Santa Maddalena or Rosso di Menfi.**

PER PORTION:
493 calories, fiber 0.48 oz/13.9 g

Cinderella's risotto

1½ cups/320 g of rice, 7 oz/200 g of diced squash,
1 bunch of chicory, cut into strips,
a few grains of fresh green pepper, 1 clove of garlic, grated,
½ a spring onion, chopped,
1 small glass of white wine,
meat stock, butter, salt and pepper,
1 small onion, chopped, parmesan to taste.

SERVES 4

WINES - **This is a simple dish with strong, rustic flavors. The wine should be a mature dry white, such as the Colli Berici Tokay or the Ansonica del Giglio.**

METHOD

1. Rapidly brown the chicory and the squash in butter with a little onion and garlic.

2. Brown the rice in butter, with onion, and sprinkle with the white wine. Add it to the chicory and squash and, adding the hot stock a little at a time, cook until soft.

3. Serve with a sprinkling of parmesan to taste.

PER PORTION:
484 calories, fiber 0.09 oz/2.6 g

Pasta twists with creamed squash

AND TRUFFLES

7 oz/200 g of fresh green pasta,
7 oz/200 g of fresh saffron pasta,
3½ oz/100 g of fresh ricotta,
3½ oz/100 g of caprino (goat's cheese),
3½ oz/100 g of crescenza (soft cheese from Lombardy),
1 lb/500 g of squash, ½ cup/50 g of parmesan,
1¾ oz/50 g of black truffle, extra virgin olive oil,
salt, pepper, nutmeg, marjoram,
1 egg, ½ cup/100 g of butter.

SERVES 4

WINES - **For this dish choose young dry white wine with a slightly herbal bouquet, such as Valle Isarco Sylvaner or Colli Martani Grechetto.**

METHOD

1. Roll out the sheets of green and saffron pasta to a thickness of ½ in/1 cm. Cut it into strips ¾ in/2 cm wide and superimpose them, alternating the colors, then pass them through a pasta maker so as to obtain a sheet of thinner pasta.

2. Mix together the ricotta, caprino, and crescenza, season them with salt, pepper, and nutmeg and leave to rest in the refrigerator for about 1 hour.

3. Cut the pasta sheet into 3 in/7 cm squares. Brush them with egg, fill them with the cheese mixture and close them by twisting together the two corners at each end.

4. Remove the rind and seeds from the squash, then boil it in salted water until soft. Mash it, beating in a little olive oil and marjoram and some of the black truffle. Keep hot.

5. Boil the pasta twists in plenty of salted water, then toss them in butter and parmesan.

6. Pour the creamed squash into a serving dish, arrange the parcels on top, and garnish with truffle and marjoram. Serve very hot.

PER PORTION:
916 calories, fiber 0.22 oz/6.4 g

Squash and radish sauce

7 oz/200 g of squash, rind and seeds removed,
7 oz/100 g of radishes,
7 tbsp/50 g of toasted squash seeds (or walnut kernels),
1 tbsp of vinegar, salt, sugar, 3 tbsp of olive oil.

SERVES 8

WINES - **Because this sauce goes equally well with boiled meats and with fish dishes, particularly salmon or trout, the wine should be chosen accordingly.**

METHOD

1. Steam the squash and sieve it. Add a dash of vinegar and a pinch of sugar and leave it to rest for ½ hour.

2. Grind the radishes in a blender, then mix in the squash seeds (or walnut kernels). Season with salt and leave to rest for ½ hour.

3. Pour the squash into a sauceboat and add the radish. Stir well, adding a little olive oil.

4. Serve at room temperature.

PER PORTION:
44 calories, fiber 0.03 oz/0.8 g

Anchovies in batter

WITH CREAMED SQUASH

16 fresh anchovies of medium size, 4 eggs,
1 cup/100 g of flour, 2 tsp of parmesan, salt, pepper, basil,
7 oz/200 g of squash, 1¼ oz/50 g of leek,
2 tsp of coriander seeds, 7 oz/200 g of celeriac,
3½ tbsp of extra virgin olive oil, coriander leaves,
fish or vegetable stock,
olive oil for frying.

SERVES 4

WINES - This dish should be accompanied a dry white wine, such as Collio Tokay or Malvasia del Carso.

PER PORTION:
470 calories, fiber 0.09 oz/2.7 g

METHOD

1. Bone the anchovies, leaving the fillets still joined at the tail. Carefully rinse and dry them.

2. Prepare the batter, beating the eggs into the flour. Season with parmesan, salt, pepper and finely chopped basil.

3. Trim the squash and cut it into regular pieces. Wash the leek and cut it into rings.

4. In a frying pan, stew the squash and the leek with a little oil over a moderate heat, cover with the fish or vegetable stock and cook until the squash is tender.

5. Put the squash and leek into a blender and whisk them up with the extra virgin olive oil.

6. Toast the coriander seeds in the oven for a few minutes, then grind them and add them to the creamed squash and leek.

7. Dip the anchovies in the batter and deep-fry them in olive oil. When they are golden brown, remove from the pan and drain on kitchen paper.

8. Trim the celeriac and slice it thinly. Stamp out little rounds with a pastry cutter. Flour them and fry until crisp.

9. Pour the creamed squash mixture onto a serving dish and arranged the fried anchovies on top, surrounded by the celeriac crisps and garnished with coriander leaves.

Grilled octopus and squash

1lb 5 oz/600 g of octopus, 7 oz/200 g of a long squash,
1¼ oz/50 g of red Acquaviva onion,
4¼ oz/120 g of celery heart, 1½ oz/40 g of black olives,
5¼ oz/150 g of borage, ½ cup of extra virgin olive oil,
⅓ cup/80 ml of lemon juice, 4 tsp/20 ml of vinegar,
salt, black pepper, parsley, chervil.

SERVES 4

WINES - **To complement the complex flavors in this dish, choose mature dry white wine, such as Torgiano or Biancolella from Ischia.**

METHOD

1. Cut the onion into fine strips and soak in cold water for around 1 hour.

2. Remove the rind and seeds from the squash and slice it. Cut the slices into triangles and grill them. Season with salt, pepper, chopped parsley, vinegar and a trickle of olive oil. Keep hot.

3. Slice the celery finely and salt it to make it lose its water. Rinse and dress it with a little lemon juice, olive oil and black pepper.

4. Stone and chop the olives.

5. To make the vinaigrette, remove the tough outer leaves from the borage and liquidize it with the olive oil and lemon juice. Season with salt and pepper and pass it through a fine strainer, reserving the liquid.

6. Take the octopus, cut it so as to separate the tentacles and grill them, taking care not to allow them to become too dry.

7. Arrange the triangles of grilled squash in a fan on each plate with the octopus on top.

8. Dress them with the borage vinaigrette and sprinkle with the marinated celery, the chopped olives and the raw red onion. Garnish with a few sprigs of chervil.

PER PORTION:
374 calories, fiber 0.07 oz/1.9 g

Squash flavored with radish

AND PRAWNS

7 oz/200 g of squash (preferably the Mantuan variety),
¾ oz/20 g of radish,
¼ cup/40 g of brown rice cooked and cooled,
8 large prawns each weighing about 3½ oz/100 g,
2½ tbsp/40 ml of olive oil, 4 tsp/20 ml of balsamic vinegar,
1 cup/100 g of flour, ⅔ cup of ice-cold water, oil,
20 leaves of tarragon.

SERVES 4

WINES - In this dish the sweetness of the squash and scampi contrast with the piquancy of the radish and the acidity of the vinegar sauce. Choose a young white wine with a slightly herbal bouquet, such as Alto Adige Riesling or Sauvignon Poggio alle Gazze.

METHOD

1. Clean the scampi, reserving the heads to make stock: crush the heads, cover them with water and simmer for 20 minutes.

2. Pass the stock through a fine sieve, place it on the heat and reduce it to ¼ cup. Allow to cool.

3. When the liquid has cooled, add 4 tsp/20 ml of olive oil, balsamic vinegar and salt.

4. Lightly salt the cleaned scampi and steam them for 2 minutes.

5. Cut the trimmed squash into thin slices and flour them.

6. Prepare a batter with the flour and the ice-cold water. Dip the slices of squash in the batter and fry them for 3 minutes. Remove them from the pan and drain on kitchen paper. Keep warm.

7. Brown the rice in a frying pan with the remaining olive oil and a pinch of salt. When it begins to sizzle add the grated radish.

8. Divide the squash between four plates. Sprinkle it with the rice and radish. Add two scampi to each plate, arranging them in a cross. Dress with the balsamic vinegar sauce and garnish with the tarragon leaves.

PER PORTION:
329 calories, fiber 0.04 oz/1.4 g

Anchovy and squash kebabs

14 oz/400 g of fresh anchovies, filleted,
7 oz/200 g of yellow squash, rind and seeds removed,
7 slices of bread, basil, salt, pepper, ½ tbsp of parsley,
½ tbsp of fresh basil leaves,
2 cloves of garlic, salt, pepper.

SERVES 4

METHOD

1. Cut the squash into 1 in/2 cm cubes. Remove the crust from 3 of the slices of bread and cut them into 1 in/2 cm squares.

2. Take some wooden skewers and on each thread a combination of the anchovies, folded in half, the squares of bread, the cubes of squash, and basil.

3. In a blender grind the remaining slices of bread with the parsley, basil, garlic, salt and pepper. Roll the kebabs in this mixture.

4. Heat a little oil in a non-stick frying pan and fry the kebabs.

5. Serve with a mixed salad dressed with lemon vinaigrette.

Note: *The success of this recipe depends to a large extent on the squash being very fresh and not too ripe. Otherwise, it is likely to become soft and watery during cooking.*

WINES - **The strong flavor of the anchovies blends deliciously with the sweetness of the squash. Choose a young dry white wine with an intense, fruity bouquet, such as Vesuvius or Bianco d'Alcamo.**

PER PORTION:
134 calories, fiber 0.03 oz/0.8 g

Monferrina cake

WITH SQUASH AND APPLES

1 lb 5 oz/600 g of yellow squash, 1 lb 5 oz/600 g of apples,
2 cups/300 g of crushed macaroons, ½ cup/100 g of sultanas
soaked in rum, ½ cup/100 g of sugar, 3 eggs,
1 cup/100 g of cocoa,
1 cup of milk, butter and breadcrumbs.

SERVES 6-8

WINES - This cake can be enjoyed with young and aromatic, possibly sparkling, wine, such as Asti Moscato or sweet Colli Piacentini Malvasia.

METHOD

1. Wash and peel the squash and the apples. Cut them into pieces, place in a pan over a low heat and cook them to a pulp.

2. When the moisture has evaporated, leave them to cool.

3. Meanwhile, preheat the oven to 355–390°F/180–200°C. Mix together all the other ingredients except the breadcrumbs. Add the cooled squash and apple pulp.

4. Butter a cake tin, sprinkle it with breadcrumbs, and pour in the mixture.

5. Bake for ½ hour, or until a knife inserted into the cake comes out clean.

PER PORTION:
317 calories, fiber 0.19 oz/5.5 g

Squash cake

WITH ANISEED

3 lb 5 oz/1.5 kg of squash, 1 cup/200 g of sugar,
6½ tbsp of cream, 6½ tbsp of milk, 4 eggs,
2 tbsp of plain white flour, 4 macaroons, crushed,
1 handful of aniseed.

SERVES 8

METHOD

1. Cut up the squash and bake it at a low temperature to extract the moisture.

2. Meanwhile mix together all the other ingredients.

3. When the squash is cooked, sieve it and combine it with the other ingredients, stirring well.

4. Raise the oven temperature to 355–390°F/180–200°C. Butter a cake tin, pour in the mixture and bake for 1 hour.

5. Allow to cool before serving.

WINES - **This cake, which carries the characteristic taste of southern Italian confectionery, should be enjoyed with sweet white wine, possibly a muscatel or other dessert wine, with a fruity, slightly spicy bouquet, such as Malvasia delle Lipari or Alto Adige Goldmuskateller.**

PER PORTION:
114 calories, fiber 0.04 oz/1.2 g

Squash delight

WITH PISTACHIO AND ALMOND PASTE

10½ oz/300 g of squash, rind and seeds removed,
7 oz/200 g of ricotta,
9 oz/250 g of pistachio and almond paste,
1 tbsp of icing sugar.

SERVES 4

METHOD

1. Bake the squash at a low temperature so as to remove the moisture, then mash it.
2. Whisk up the ricotta with the icing sugar and ⅔ of the squash.
3. Thinly roll out the pistachio and almond paste and cut out some disks 4 in/10 cm across.
4. Spoon some of the mashed squash into individual dishes, drop a spoonful of the ricotta mixture into the center, cover it with a disk of almond and pistachio paste, and serve.

WINES - A suitable accompaniment for this very sweet dessert, strongly flavored by the pistachio and almonds, would be a muscatel or other sweet white wine, such as the Moscato di Pantelleria or the Torchiato di Fregona.

PER PORTION:
333 calories, fiber 0.13 oz/3.8 g

Rice and squash flan

For the shortcrust pastry:
½ cup of butter, ½ cup/100 g of icing sugar,
the yolks of 2 eggs, 2 cups/240 g of flour,
the contents of 1 vanilla bean, salt,
1 tsp of baking powder.

For the filling:
1¾ oz/50 g of squash, rind and seeds removed and diced,
¼ cup/50 g of rice, ¾ cup of water,
6½ tbsp of confectioner's custard, the whites of 3 eggs,
2 tbsp/30 g of sugar.

SERVES 5–6

WINES - **For this dessert, choose a sweet white wine, possibly a muscatel, such as Recioto di Soave Classico, Colli Orientali del Friuli Piccolit or Malvasia di Bosco.**

METHOD

1. Soften the butter, mix it with the sugar, the vanilla, the salt and the egg yolks and stir well. Then add the flour and baking powder. Gently press the pastry into a ball, wrap it in clingfilm and let it rest in the refrigerator for ½ hour.

2. Put the water, rice and squash into a small pan, bring to a boil and allow to simmer until the rice is cooked, adding more water if necessary.

3. Add the custard to the rice and squash mixture, stirring it in well.

4. Whisk the egg whites with the sugar until stiff and fold into the rice and squash mixture.

5. Heat the oven to 355°F/180°C. Roll out the pastry and line a flan case with it. Pour in the mixture and bake for 40 minutes.

PER PORTION:
483 calories, fiber 0.04 oz/1.1 g

Squash molds

WITH STAR ANISEED CUSTARD

For the molds:

1 lb 5 oz/600 g of squash, rind and seeds removed,
4 cups/1 l of milk, 1 cup/200 g of sugar,
2 tbsp of Amaretto liqueur, the contents of 1 vanilla bean,
3 eggs.

For the custard:

the yolks of 3 eggs, 1 cup of milk, ⅓ cup/80 g of sugar,
3 star aniseed.

For the garnish:

a handful of diced squash, 1 tbsp of sugar.

SERVES 6

WINES - **This pleasantly sweet dessert, with its delightful flavor of Amaretto the vanilla, should be accompanied by a delicate, young, sweet white wine such as Colli Orientali del Friuli Ramandolo or Albana di Romagna.**

METHOD

1. Put all the ingredients for the molds, except the eggs, into a saucepan and simmer for about 40 minutes. Beat the eggs, whisk them into the mixture and pass it through a fine sieve.

2. Pour the sieved mixture into individual molds. Place the molds in a dish of water and bake at 250°F/120°C for about 50 minutes.

3. Meanwhile prepare the custard. Beat the egg yolks with the sugar. Add the aniseed to the milk and bring it to a boil. Pour it through a sieve into the egg mixture, stirring all the time. Return the liquid to the heat but without letting it boil. Leave to cool.

4. Spoon a little of the sauce onto individual dishes and serve the molds warm, turning them out onto the sauce, surrounded by diced squash tossed in the frying pan and caramelized with the sugar.

PER PORTION:
428 calories, fiber 0.02 oz/0.5 g

Pears stuffed with squash

6 cooking pears, 4 cups/1 l of good red wine,
1½ cups/400 g of sugar, the yolks of 4 eggs,
1 lb 6 oz/600 g of squash,
7 oz/200 g of macaroons, 6 kept whole and the rest crushed,
3 tbsp/20 g of pine kernels, coarsely chopped,
2 tbsp/20 g of sultanas,
ground cinnamon, liqueur.

SERVES 6

WINES - This dessert demands a sweet red wine, possibly sparkling, such as Brachetto d'Acqui or Alto Adige Moscato Rosa.

METHOD

1. Flavor the wine with sugar and cinnamon and boil the pears in it for about 1 hour. Leave them to cool, then remove the cores, reserving them and leaving the fruit as whole as possible.

2. Meanwhile bake the squash at 300°F/150°C for about 1½ hours, sprinkling it occasionally with granulated sugar and liqueur to taste.

3. When the squash is cooked, pass it through a sieve. Sieve the pear cores. Add the sieved squash and pear to the mixture together with the sultanas, the crushed macaroons, the pine kernels, a little cinnamon, and more sugar if desired.

4. Bind the mixture with the egg yolks and spoon it into the cavity in the pears. Dip the 6 whole macaroons in liqueur and place one on the top of each pear. Glaze the pears with the syrup obtained from their cooking.

PER PORTION:
383 calories, fiber 0.18 oz/5.1 g

Squash soufflé

10½ oz/300 g of squash, rind and seeds removed,
½ cup/150 g of sugar, 1 g of vanilla, 4 eggs, butter and sugar.

Serves 4

METHOD

1. Steam the squash, sieve it and add the sugar and vanilla.

2. Bind this purée with the egg yolks.

3. Whisk the egg whites until stiff and fold well into the mixture.

4. Butter and sugar 4 individual molds and fill them.

5. Heat the oven to 355°F/180°C and bake the soufflés for 20 minutes.

6. Serve immediately.

Wines - **For this delicate dessert choose a sweet white wine, such as Nus Malvoisie Fletri or Trentino Vino Santo.**

PER PORTION:
123 calories, fiber 0.01 oz/0.4 g

Squash and almond mousse

7 oz/200 g of squash, the yolks of 5 eggs,
1 cup/200 g of sugar, 1⅔ cups/400 ml of cream, whipped,
the whites of 2 eggs, vanilla flavoring, powdered cinnamon,
1⅔ cups/400 ml of custard (made from the yolks of 3 eggs,
4 tbsp sugar and 1 cup of milk),
15 oz/420 g of macaroons, crushed and sieved,
1 cup/100 g of almonds, sliced and toasted.

SERVES 8

WINES - This richly flavored mousse should be accompanied by a white dessert wine with an aromatic bouquet, such as Moscato di Loazzolo or Albana di Romagna Passita.

1. Bake the squash at 355°F/180°C for 20–25 minutes then pass it through a sieve.

2. Whisk the egg yolks with the sugar and add a few drops of vanilla flavoring.

3. Whisk the egg whites until stiff.

4. Mix the squash with the yolks thoroughly using a wooden spatula. Add the cream, followed by the egg whites.

5. Pour the mixture into 8 individual molds and leave them in the icebox for at least 4 hours.

6. To prepare the custard, whisk the egg yolks with the sugar, heat the milk to boiling point, and add to the yolks and sugar. Stir thoroughly and return to the heat but do not allow to boil. Add the macaroons.

7. Spoon the custard onto individual plates and turn out each mold on top. Sprinkle with the almonds and cinnamon powder.

PER PORTION:
582 calories, fiber 0.08 oz/2.4 g

Squash jam flan

For the squash jam:

1 squash weighing about 11 lbs/5 kg,
3 cups/800 g of granulated sugar, ¼ cup/200 g of honey,
½ cup/100 g of grated orange and citron zest,
the grated zest of 1 lemon,
6½ tbsp of Marsala or white wine.

For the puff pastry:

2½ cups/300 g of plain white flour, 10 tbsp/150 g of butter,
½ cup/100 g of sugar, the yolks of 2 eggs,
grated lemon zest, salt.

SERVES 8

Wines - A suitable accompaniment for this succulent dessert with a spicy aroma is a sweet wine such as Colli Orientali del Friuli Picolit or Muffato della Sala.

METHOD

1. Cut and slice the squash then soak it overnight in a mixture of the sugar and the Marsala.

2. Next day, add the honey and the orange, citron, and lemon zest.

3. Heat until the mixture attains the consistency of jam.

4. To make the pastry, cream butter and sugar, add the egg yolks, then the flour, lemon zest, and a pinch of salt. Mix well and leave to rest in the refrigerator for 1 day.

5. Line a flan tin with the pastry. Spoon in the squash jam and decorate with strips made from pastry offcuts.

6. Heat the oven to 390°F/200°C and bake the flan for about 20 minutes. Serve with a sprinkling of icing sugar.

PER PORTION:

576 calories, fiber 0.09 oz/2.7 g

Sweet squash and almond flan

WITH MACAROON SAUCE

For the flan:

1¼ oz/50 g of squash, rind and seeds removed,
1 oz/30 g of almonds, 4 tsp of milk, ¼ cup/60 g of sugar,
3½ tbsp of butter, ½ cup/50 g of flour, 3 eggs, beaten,
¼ cup/30 g of breadcrumbs, 1 oz/25 g of macaroons.

For the macaroon sauce:

4 tsp/20 ml of milk, ½ cup/100 g of sugar, 2 tsp of cornflour,
the yolks of 3 eggs, 2½ tbsp of cream, Amaretto liqueur.

For the garnish:

¾ oz/20 g of macaroons, crushed, 1 oz/25 g of butter,
3½ oz/100 g of squash, rind and seeds removed,
3½ oz/100 g of whole fresh coconut, 5 tsp of Amaretto
liqueur, icing sugar, fresh mint leaves, powdered cinnamon.

SERVES 4

WINES - **This interesting dessert, with an array of flavors, should be accompanied by a sweet white wine or dessert wine, such as the Erbaluce di Caluso Passito, the Nettare dei Santi Passito, the Moscadello di Montalcino or the Malvasia delle Lipari.**

METHOD

1. Finely dice the squash and shred the almonds (unskinned) lengthwise. Put both in a saucepan with the milk and sugar and bring to a boil.

2. Separately heat the butter and stir in the flour. Add to the squash, almonds and milk so as to make a roux. Cool the base of the pan and add the eggs.

3. Mix the breadcrumbs with the crushed macaroons. Butter four individual molds and sprinkle them with the macaroon and breadcrumb mixture. Fill with the squash mixture, place the molds in a large dish filled with water and bake at 300°F /150°C for about 1 hour.

4. To make the macaroon sauce, beat the egg yolks and sugar together, add the cornflour, followed by the milk. Heat gently and do not allow to boil. Allow to cool. Whip the cream and stir it into the custard together with a little Amaretto liqueur.

5. For the garnish, cut the squash and coconut into strips, put them into a frying pan with the butter and a little sugar, and when they start to caramelize bathe them with the Amaretto liqueur.

6. Serve the flan hot with the macaroon sauce and the hot squash and coconut garnish. Decorate with a few mint leaves and dust the flan with icing sugar and cinnamon.

PER PORTION:
438 calories, fiber 0.65 oz/1.4 g

Acknowledgments

RISTORANTE
AL BERSAGLIERE
Roberto Ferrari
Via Goitese 260
46044 GOITO (MN)

RISTORANTE SADLER
Claudio Sadler
Via Troilo 14
20136 MILANO

RISTORANTE LA FRASCA
Marco Cavallucci
Viale Matteotti 34
47011 CASTROCARO TERME (FO)

RISTORANTE L'ALBERETA
Gualtiero Marchesi
Via Vittorio Emanuele 11
25030 ERBUSCO (BS)

RISTORANTE DA GIGETTO
Luigi Bortolini
Via A. De Gasperi 4
31050 TREVISO

TRATTORIA IL FOCOLARE
Agostino D'Ambra e Rosario Sgambati
80074 CASAMICCIOLA TERME (NA)

RISTORANTE LA TAVOLA D'ORO
Giovanna Gasparello
Via Santa Chiara 2
31100 TREVISO

RISTORANTE IL CASCINALE NUOVO
Walter Ferretto
Statale Asti-Alba, 15
14057 ISOLA D'ASTI (AT)

RISTORANTE TIVOLI
Walter Bianconi
Loc. Lacedel
32043 CORTINA D'AMPEZZO (BL)

RISTORANTE BISTROT CLARIDGE
Vincenzo Cammerucci
Via dei Mille 55
47042 CESENATICO (FO)

RISTORANTE GABBIA D'ORO
Domenico Burato
Loc. Gabbia
37063 ISOLA DELLA SCALA (VR)

LA LOCANDA DELLA TAMERICE
Igles Corelli
Via Argine Mezzano 2
44020 OSTELLATO (FE)

RISTORANTE VECCHIO MULINO
Luca Bolfo
Via Monumento 5
27012 Certosa di Pavia (PV)

RISTORANTE AIMO E NADIA
Aimo e Nadia Moroni
Via Montecuccoli 6
20147 MILANO

RISTORANTE CUCINA DEL MUSEO
Alberto Vaccari
Via Sant'Agostino 7
41100 MODENA

RISTORANTE CASA FONTANA
Roberto Fontana
Piazza Carbonari 5
20125 MILANO

CASANOVA GRILL HOTEL PALACE
Umberto Vezzoli
Piazza della Repubblica 20
20124 MILANO

TRATTORIA IL MOLINETTO
Stefano Gandini
Loc. Molinetto
42033 CARPINETI (RE)

ACKNOWLEDGMENTS

RISTORANTE
GALLERIA DELL'HOTEL
PRINCIPE DI SAVOIA
Romano Resen
Piazza della Repubblica 17
20124 MILANO

LA LOCANDA DI ALIA
Pinuccio Alia
Contrada Jetticelle
87012 CASTROVILLARI (CS)

RISTORANTE
LE COLLINE CIOCIARE
Salvatore Tassa
Via Prenestina 27
03010 ACUTO (FR)

RISTORANTE
AMBASCIATA
Romano Tamani
Via Martiri di Belfiore 33
46026 QUISTELLO (MN)

ANTICA OSTERIA DEL PONTE
Ezio Santin
Piazza Negri 9
20080 CASSINETTA DI LUGAGNANO
(MI)

RISTORANTE
DAL PESCATORE
Nadia Santini
Loc. Runate
46013 CANNETO SULL'OGLIO (MN)

RISTORANTE
BORGO ANTICO
Sergio Cantatore
Piazza Municipio 20
70056 MOLFETTA (BA)

SYMPOSIUM
QUATTRO STAGIONI
Lucio Pompili
Via Cartoceto 38
61030 CARTOCETO (PS)

RISTORANTE
PAOLO TEVERINI
Paolo Teverini
Piazza Dante 2
47021 BAGNO DI ROMAGNA (FO)

Acknowledgments

GRAND HOTEL
VILLA ROMANAZZI-CARDUCCI
Antonio De Rosa
Via Capruzzi 326
70124 BARI

GOLF HOTEL RIVA DEI TESSALI
Giovanni Maggi e Virgilio Corrado
74025 MARINA DI GINOSA (TA)

RISTORANTE JOIA
Pietro Leemann
Via P. Castaldi 18
20124 MILANO

Laura NICCOLAI
Via Termine 9
80064 S. AGATA SUI DUE GOLFI (NA)

RISTORANTE
LA CONTEA
Claudia e Tonino Verro
Piazza Cocito 8
12052 NEIVE (CN)

HOTEL DELLA POSTA
Renato Sozzani
Piazza Garibaldi 19
23100 SONDRIO

DON ALFONSO 1890
Alfonso Iaccarino
C.so S. Agata 11
80064 S. Agata sui due Golfi (NA)

L'ANTICA ARTE DEL DOLCE
Ernst Knam
Via Anfossi 10
20135 MILANO

RISTORANTE ITALIA
Sergio Carboni
Via Garibaldi 1
26038 TORRE DE' PICENARDI (CR)

RISTORANTE FERRANDO
Piero Ferrando
Via D. Carli 110
16010 S. CIPRIANO DI SERRA RICCO'
(GE)

FORESTERIA CREDITO ITALIANO
Fred Beneduce
MILANO

RISTORANTE ALBERGO DEL SOLE
Franco Colombani
Via Mons. Trabattoni 22
20076 MALEO (LO)

ARTE DELLA PASTICCERIA
Enrico Parassina e Daniele Allegro
Via A. Diaz 7
35031 ABANO TERME (PD)

Ristorante LANCELLOTTI
Angelo Lancellotti
Via Grandi 120
41019 SOLIERA (MO)

THE FOLLOWING ARE THANKED FOR THEIR
INVALUABLE CONTRIBUTION:
*Biblioteca Internazionale di Gastronomia,
Eugenio Medagliani, Famiglia Nizzoli,
Attilio Pollastri, Paola Salvatori,
Noemi Govi, Laura Valastro.*

THE CALORIFIC VALUES AND FIBER CONTENT
WERE CALCULATED WITH THE AID OF THE
PROGRAM:
*Mangi 3, developed by 3GTO Software
and Marco Riva*